SOLDERING AND BRAZING

A Handbook for Mechanics giving Directions for Soft Soldering, Hard Soldering and Brazing.

Contains Data on Preparing Various Solders and Includes a Multitude of Soldering Kinks. Gives Design and Construction of Electric Soldering Coppers and Heating Devices

BY

RAYMOND FRANCIS YATES

EDITOR OF "EVERYDAY ENGINEERING"

AUTHOR OF "MODEL MAKING," "WORKSHOP PRACTICE FOR HOME MECHANICS," ETC.

A Practical Hand-book for Everyone Interested in the Process of Soldering and Brazing

FULLY ILLUSTRATED

THE NORMAN W. HENLEY PUBLISHING CO.,
2 WEST 45TH STREET
NEW YORK
1920

PRINTED IN U. S. A.

Composition, Electrotyping and Press Work
By Publishers Printing Company, New York

PREFACE

Soldering is a much used process, yet few are able to manipulate the soldering copper successfully. The author has watched many home mechanics trying to mend articles by the application of solder. It is amusing to see a would-be mechanic ambitiously trying to mend a pot or kettle with the soldering rolling in little balls as fast as it is applied. When solder does not "stick" something is missing—something is wrong.

It is known that a little experience is the greatest aid to soldering, but, on the other hand, the rankest amateur can apply solder so that it will stay in place if he will inform himself regarding the fundamentals of the process, which are few and very simple.

This little book has been prepared for mechanics, "tinkers" and everyone that has occasion to apply soldering either as a part of their daily work or part of their hobby of "making things." The subject-matter has been presented in a non-technical way, absolutely devoid of confusing statements or dry theory. The author has tried to include everything that will assist the beginner in accomplishing successful work.

This little book is launched into the great sea of mechanical literature with the hope that it will find its way into the hands of everyone who has had occasion to find out the sad truth that they are unable to make solder "stick."

RAYMOND FRANCIS YATES,
New York City.

February, 1920.

TABLE OF CONTENTS

PART I

SOFT SOLDERING

PART II

HARD SOLDERING AND BRAZING

PART III

BRAZING

PART IV

HEATING DEVICES

PART V

SOLDERING NOTES

Soldering and Brazing

PART I

SOFT SOLDERING

Soft soldering is the simple process of uniting two pieces of metal by means of an alloy with a low melting point. It is not a difficult process, but, like many others, it is one that requires a certain "knack." A man may be well equipped with the fundamentals of simple soft soldering as far as knowledge goes, but on the other

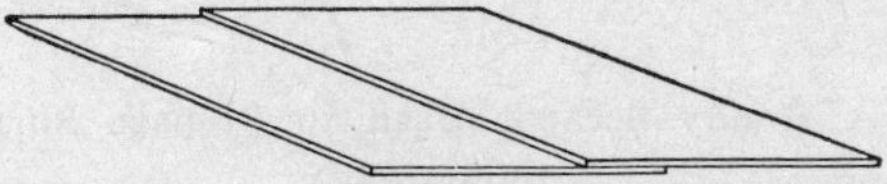

Fig. 1.—Two Pieces of Tin to be Joined by Soldering.

hand he may not be able to solder successfully. It seems to be one of those processes that requires just a certain amount of experience before it can be done successfully. The beginner's trouble is in the solder not "sticking." When he does get it to "stick" it generally adheres in lumps and makes a very unsightly job.

In a good soldering job, the solder is spread or "flowed" evenly over the metallic surfaces to which it is applied. Solder applied in this way will adhere until it is removed by the further application of heat. Solder that is applied in lumps, can generally be picked off with the fingers.

The process of soft soldering will be described progressively. It will be assumed that two oblong pieces of sheet tin are to be joined together by soft soldering. The pieces and the joint to be made are shown in Fig. 1. The first thing necessary is the cleaning of surfaces where the soldering alloy is to be applied. This is where many mechanics make a mistake. Successful soldering depends largely upon the proper preparation of the surfaces. If the surfaces are not prepared properly,

Fig. 2.—A Handy Scraper Used to Prepare Surfaces for Soldering.

the greatest difficulty will be had in making the solder adhere.

A fine grade of emery or carborundum cloth is very suitable for use in preparing metallic surfaces for soldering. A few strokes of the paper will bring the metal to a clean, semi-polished condition with the removal of all foreign matter. The little scraper illustrated in Fig. 2 will be found useful in cleaning metal surfaces. The triangular portion is made of a piece of hardened steel. The scraping edges of the device are ground sharp. This simple little tool is easy to make and it can be used on many occasions.

After the metallic surfaces to be joined are properly cleaned, the workman should refrain from touching them with his fingers, as this will leave a deposit of grease that will interfere with the solder adhering. The surfaces should not be left exposed to the atmosphere any longer than is necessary after they are thoroughly cleaned. The object of cleaning the surfaces is to remove the thin film of oxide that always forms on the surface of the common metals. This thin film of oxide would entirely prevent the soldering from adhering if it was not removed. This film of oxide forms more rapidly on some metals than it does on others. In the case of aluminum, it forms instantaneously. Therefore, it is quite impossible to solder aluminum by any of the common methods of applying solder.

After the surfaces of the two pieces of tin have been thoroughly cleaned by the process described above, a

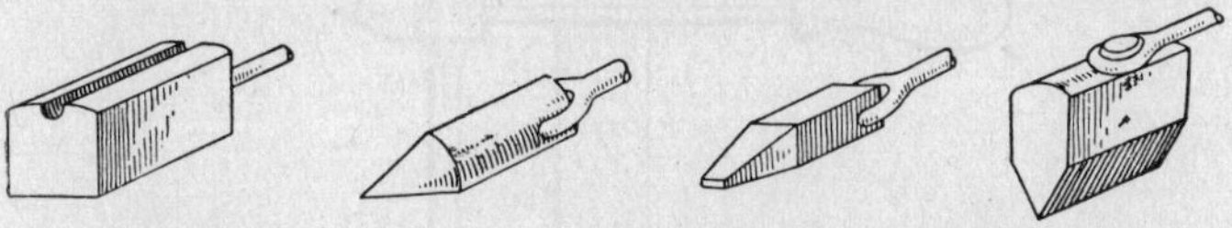

Fig. 3.—Different Shaped Soldering Coppers.

soldering flux should be applied to them. This flux will prevent further oxidation of the surfaces and will cause the solder to adhere without further trouble. There are many different soldering fluxes available and some of these will be described in detail in a later part of this Chapter. For the time being, rosin only will be considered. When rosin is used as a flux it should be crushed to a powder and applied with a stiff brush. We now have the surfaces prepared and ready to receive the molten solder.

Solder is applied with what is known as a soldering copper. This device may be more commonly known as a soldering "iron" but the writer would prefer to use the word copper.

Soldering coppers are made in different sizes and weights. A number of different shapes of soldering coppers are shown in Fig. 3. Each one of these coppers has a certain use. The one shown at A is the type used for all common soldering jobs. Soldering coppers selected for ordinary work should be of medium weight. The weight of the soldering copper is a very important consideration. The heavy copper once heated will hold its heat longer than a small one. On the other hand, a real heavy soldering copper is more difficult to manipulate than a medium or light-weight one. Beginners should start by using a medium-weight soldering copper. For small, delicate jobs, however, it will be necessary to employ a very small copper. In delicate work the

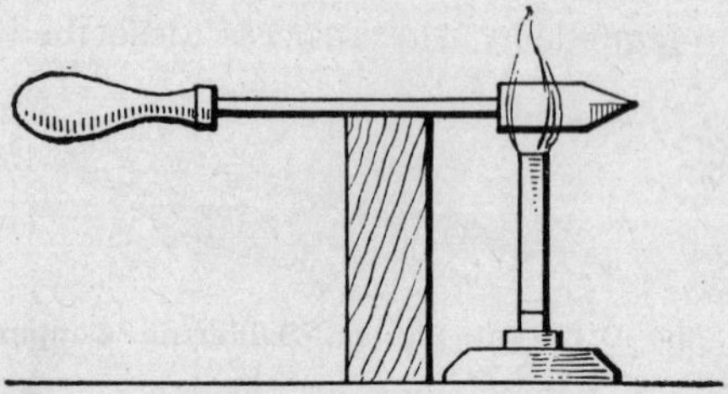

Fig. 4.—How the Soldering Copper Should be Placed for Heating.

volume of solder needed is small and, therefore, a small copper is able to meet the needs before most of its useful heat is radiated into the atmosphere.

Soldering coppers are brought to the proper temperature by several different methods. A hot gas-flame, such as that given by a Bunsen burner, makes a very suitable method of heating and, in case one of these devices is used, the soldering copper should be placed as shown in Fig. 4. It will be seen here that the point of the copper is not placed in the flame. The heat is applied at the back of the copper and, owing to the high thermal conductivity of the copper, the point will be heated to practically the same temperature as the back. The object of heating the copper in this way is to avoid burning the "tinning" off. The tinning is really solder that is applied to the tip of the copper and successful soldering cannot be accomplished without this. The process of tinning will be considered later on in this Chapter.

Gasoline blow torches are suitable for heating soldering coppers, but they produce a bad odor and an annoying roar. However, they give a perfect heat and are able to bring the soldering copper to the proper temperature within a very few minutes.

Heating should never be carried on by putting a copper in a coal fire. This would burn the tinning off

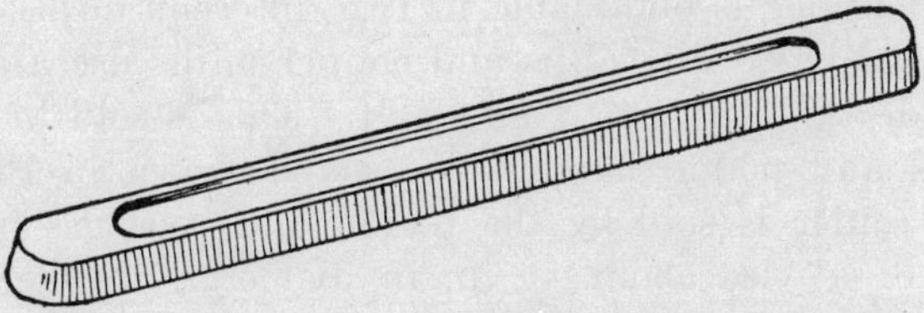

Fig. 5.—Solder is Sometimes Sold in Pound Bars, as Illustrated.

and render the device unsuitable for further use. A number of years ago tinsmiths used a portable fire pot by means of which they heated their soldering coppers with a charcoal fire. This method is obsolete to-day and the more modern gasoline torch is always used.

To solder successfully, the copper must be of just the proper temperature. It must be neither too hot nor too cold. One condition is just as bad as the other. If the copper is chilled the solder will not run freely, and very little of it will leave the point of the copper after melting. A chilled copper will not be able to communicate enough heat to the work and, therefore, the solder that does reach it will not adhere. If the soldering copper is too hot, the solder will have a tendency to form into little globules and run off the work. The solder is brought to such a high temperature that it requires some time to cool.

It is difficult to explain in writing just the proper temperature that the copper should be brought to before use. Probably the following rule covers the situation about as well as anything that can be said: The copper should be just hot enough to melt the solder after it has been in contact for a few seconds. If the solder melts instantaneously upon coming in contact with the copper it signifies that too much heat is available. On the other hand, if it is necessary to hold the solder to the copper for any great length of time the copper is not at the proper temperature; it is too cold.

Soft solder is obtainable in two different forms, i. e., wire and bar. Tinsmiths and coppersmiths use a grade of solder known as half and half. This is sold in half-pound and pound bars in the shape shown in Fig. 5. Wire solder is sold by the pound and it comes in the form of a wire about 1/8 in. in diameter. Solder was ordinarily prepared in this way for use in electrical work. Wire solder has been found very convenient, for it is used in many other ways and has done much to replace the clumsy solder bar, which requires some skill.

All the necessary operations to prepare for soldering have now been covered and the process of soldering the two pieces of tin together will now be described. Before finally applying the solder see that two pieces of tin are brought as close together as possible. The closer their edges are together the stronger the resulting joint will be. The beginner is very apt to believe that the more solder he flows on a joint the stronger it will be. This is a sorry mistake and is just opposite to the truth. The pieces of tin should be laid upon a perfectly flat surface before the solder is applied.

If a bar of solder is used, it will be necessary to take the solder from it as shown in Fig. 6. If the copper is too hot it will be found very difficult, if not impossible, to carry any solder from the bar to the work by means

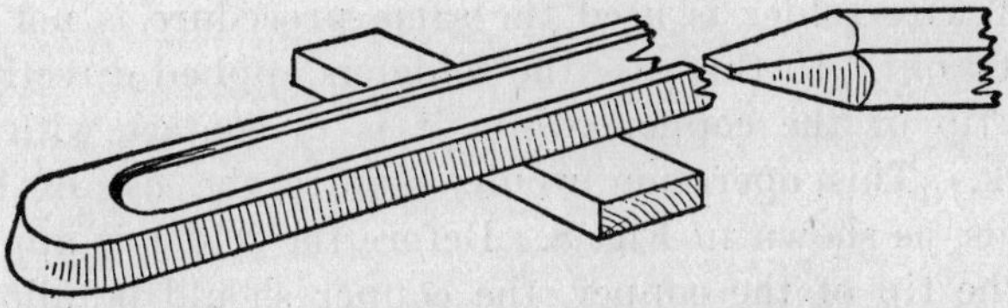

Fig. 6.—Taking Solder From a Bar With a Hot Copper.

of the copper. The copper should be just hot enough so that the solder will adhere to it without rolling off. The tip of the copper should then be placed at one end of the joint, as shown in Fig. 7. It should be allowed to remain in this position for a few seconds until sufficient heat has been communicated to the piece to be soldered.

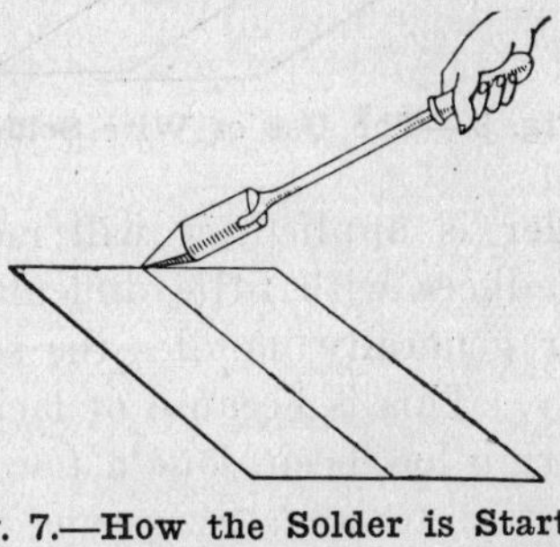

Fig. 7.—How the Solder is Started.

It must be remembered that solder will not adhere to cold surfaces and that the necessary heat must be obtained from the copper. After the copper has been held to the surface of the joint for a few seconds, the solder will leave the copper and flow onto the work, providing the copper has been properly tinned before being used.

Once the solder starts to flow, the copper should be moved slowly along the joint until the supply of solder on its tip is exhausted. The supply of solder should then be replenished and the soldering continued at the point where the supply became exhausted.

If wire solder is used the same procedure is not followed out. In this case the solder is applied directly to the tip of the copper, while it is in contact with the work. This operation would require the use of both hands, as shown in Fig. 8. Before the solder is applied to the tip of the copper, the copper should be allowed to remain in contact with the work for a few seconds.

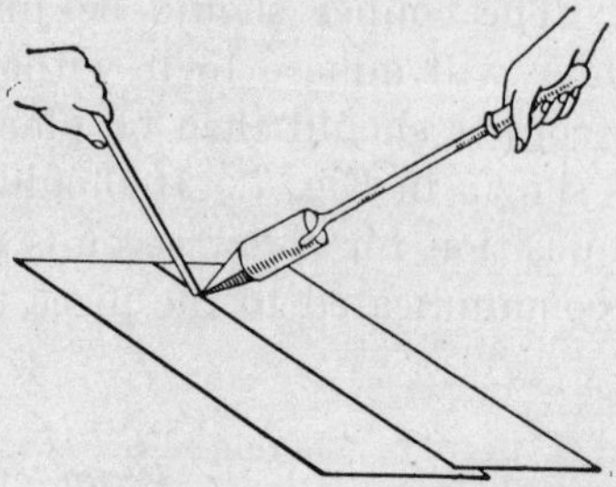

Fig. 8.—The Use of Wire Solder.

When the solder is applied, it will melt and flow to the work and adhere with little difficulty.

The beginner generally handles the soldering copper very awkwardly. This is because of lack of experience, but after soldering has been done a few times this awkwardness will start to wear away and before long the mechanic will be able to handle it like a veteran. It is positively humorous to watch some beginners trying to solder. In most cases they either apply the solder in lumps or not at all. If there is one process that exhausts the patience of the beginner it is that of soldering. If the procedure outlined in the preceding paragraphs is carefully followed and a little patience is used, the results of the first attempts should be encouraging. There is one rule the beginner should keep in mind: The soldering copper should never be permitted to become too hot. In the event that it does become too hot, it can be cooled to the proper temperature by merely stroking it across a wet rag two or three times. Overheating not only makes the process of applying the solder difficult, but greatly shortens the life of the copper. Copper when heated has a tendency to oxidize and this process of oxidization is accelerated by more heat. The higher the heat the more rapidly the copper will oxidize.

The process of tinning the soldering copper will now be considered. For this purpose a soft, yellow brick

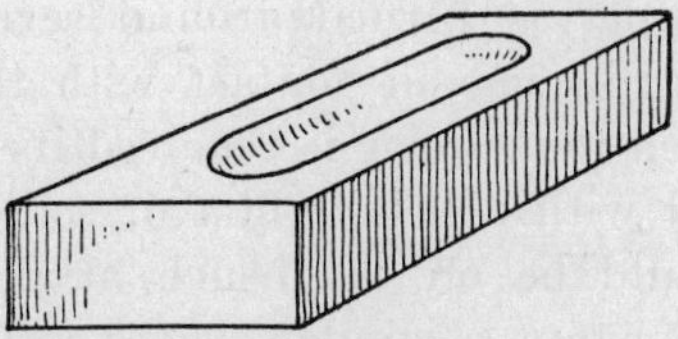

Fig. 9.—A Brick Used in Tinning a Soldering Copper.

should be obtained. A hard brick will not be suitable. A little cavity about one-half inch deep is chipped out of the brick with a cold-chisel, as shown in Fig. 9. The bottom of the cavity should be made as smooth as possible. A little rosin is then melted and poured into the cavity. It should be filled to within about $1/8$ inch of the top. Rosin can be purchased at a local hardware store for a few cents a pound. After the molten rosin is poured in the cavity a few drops of solder are put in. The tinning brick is then ready for use.

Before tinning the copper its point should be carefully trimmed up with a sharp file. The flat surfaces should be maintained and the edges made sharp and square with one another. A nicely trimmed point of a soldering copper contributes greatly toward successful soldering. The copper surface should be filed until perfectly bright. This can be done without the removing of very much copper providing the device is in good condition.

Having cleaned the point of the copper it is now heated to the proper soldering temperature. When the mechanic thinks the proper temperature has been reached he should apply the copper to the bar of solder to see if it is either too hot or too cold. If just right, the point is then placed in the cavity and the tinning brick and the copper are then stroked back and forth fifteen or twenty times. The abrasion of the copper against the brick should keep its surface clean while the solder is made to adhere to the point. During this process the copper is turned around several times so that all sides will come in contact with the bottom of the brick. When the copper is removed its point should be well covered with a bright film of solder. The tinning brick should be on the bench at all times. Its presence is a constant reminder that a well-tinned copper is a great aid to good soldering. Of course, the rosin and solder in the cavity of the tinning brick will cool and harden, but a few minutes contact with the hot copper will bring them to a fluid condition.

SOLDERING FLUXES

It has been mentioned before that fluxes are applied to retard the process of oxidation until the solder has been applied. Any metallic surface that has been oxidized cannot be soldered.

There are a number of different soldering fluxes. Some of these can be prepared by the mechanic and some can be purchased on the market. Many different brands sold under various trade names can be had.

Some of these are good and some are quite worthless. The writer feels that it is advisable to refrain from commenting on the merits of the various commercial fluxes. One thing, however, can be said with certainty: One flux is not suitable for soldering all the different metals that can be joined by this process.

Rosin has been mentioned heretofore as a suitable flux. In fact, rosin is probably one of the oldest fluxes used. It is prepared by merely grinding or crushing the lumps as they come from the dealer. A little brush used in applying rosin is shown in Fig. 10. It is made by placing a piece of woolen cloth in a metal tube and pinching the tube together. The rosin powder will ad-

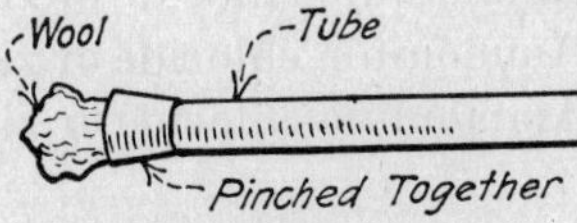

Fig. 10.—A Little Brush Used in Applying Rosin.

here to the brush when it is placed in the dish containing it. The brush it then rubbed over the work. Liquid rosin flux can be made by dissolving rosin powder in alcohol. The liquid is then applied to the work to be soldered with a small brush. When exposed to the atmosphere the alcohol evaporates, leaving a thin film of rosin on the surface of the metal.

Wire solder can be purchased that contains its own flux. The flux forms the core or center of the wire and as the solder melts, the flux melts with it and flows with the solder. Such solder wire is used largely in electrical work. Most mechanics prefer to apply the flux in the conventional way.

It must be understood that rosin is now used as a flux for soldering all metals. In fact it is used in but a few cases, as will be evident upon examining the table given below which gives the various fluxes used with the different metals:

Metal	*Flux*
Bismuth	Zinc chloride
Silver	Zinc chloride
Gold	Zinc chloride
Zinc	Hydrochloric acid
Pewter	Gallipoli oil
Steel and iron	Ammonium chloride
Galvanized steel	Hydrochloric acid
Tinned steel	Rosin or zinc chloride
Lead	Tallow or rosin
Tin	Zinc chloride or rosin
Gun metal	Ammonium chloride or zinc chloride
Copper	Ammonium chloride or zinc chloride
Brass	Ammonium chloride or zinc chloride

The zinc chloride mentioned can be prepared by the mechanic. A few pieces of zinc and a little hydrochloric acid is all that is needed. The hydrochloric acid is placed in a wide mouthed bottle and a few pieces of zinc is thrown into it. A vigorous chemical reaction will then result. When this has subsided, the preparation is ready for use. It can be applied with a small brush. It should not be allowed to touch the skin as the free acid on it will cause a burn.

Ammonium chloride, like rosin, is applied in the form of powder. It can be purchased at the corner drug store for a few cents a pound. A half pound of the substance will last ages. It is commonly known as sal ammoniac.

When soldering electrical wires a flux should be used that will not corrode the copper. If corrosion takes place after the wires have been soldered, the electrical resistance is increased to a dangerous point. A flux prepared according to the following formula will be found very satisfactory for electrical work.

Alcohol	33 parts
Glycerine	25 "
Zinc chloride	42 "

The zinc chloride crystals are dissolved in the alcohol. The glycerine is then added. The flux is applied in the usual way.

Rosin is a very good flux for soldering electric wires. This is especially true of small, delicate wires. There is absolutely no danger of corrosion when this substance is employed.

SOLDER ALLOYS

The ordinary solder bar is called "half-and-half" because it is made up of 50% tin and 50% lead. This melts in the neighborhood of 425° Fahr. The melting point of any lead-tin soldering composition depends entirely upon the percentage of tin present. If the metals used are absolutely pure, the melting point of any soldering alloy can be approximated very closely if the percentage of tin is known. The curve shown in Fig. 11 shows the melting point of solder with various percentages of tin. Here it will be seen that as the proportion of tin increases, the melting point decreases up to 70%. At this point, the curve raises again as will be seen. The following table gives the melting points of all the common soldering mixtures of lead and tin.

Lead:	100	90	80	70	60	50	40	30	20	10	0
Tin:	0	10	20	30	40	50	60	70	80	90	100
Melting Point, Fahr.:	619	560	529	504	464	428	374	365	392	421	450

The mechanic can easily prepare his own soldering mixtures by following the above table of percentages. In this way he can prepare a solder that will melt at

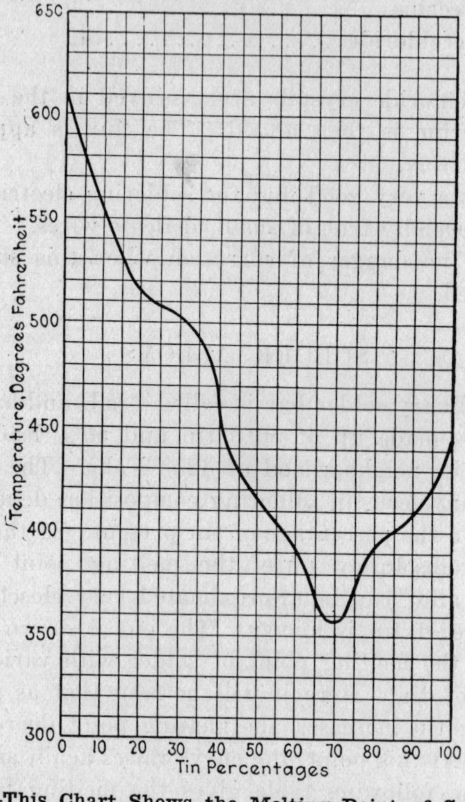

Fig. 11.—This Chart Shows the Melting Point of Solder With Various Percentages of Tin.

practically any point within 450 and 619° Fahr. The tin should be melted first in a clay crucible. When the tin has been brought to a molten condition, the lead is added. The mass is then thoroughly agitated to insure a perfectly homogeneous mixture. If this is not done, the resulting alloy will not melt at any definite point. Its melting point will depend entirely upon the percentage of tin in the part where the heat is applied. When the metals are in a thoroughly molten condition after having been well agitated, the resulting alloy can be poured into a small mould to cool. The mould is made of iron and should be about one-half inch deep and ten inches long. In the case of the small mechanic, one mould would be sufficient.

ELECTRIC SOLDERING COPPERS

Electric soldering coppers are a great convenience if the necessary current is available to operate them. It is only necessary to connect them to a 110 volt source of A. C. or D. C. current. After being connected for

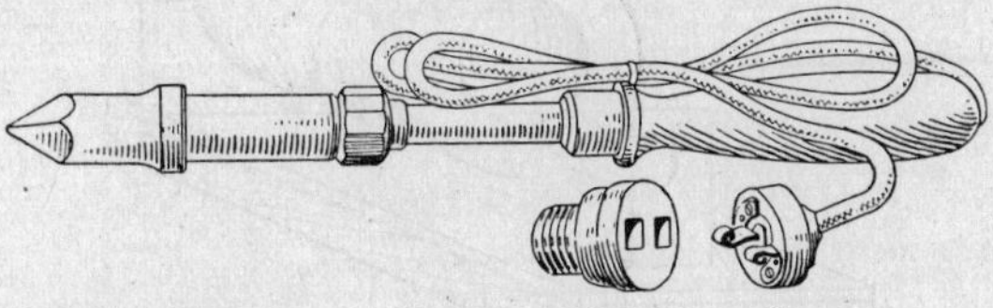

Fig. 12.—An Electric Soldering Copper.

a few minutes they reach operating temperature and this can be maintained as long as the copper is left in the circuit. If the copper becomes too hot for practical use, the current can be turned off for a few minutes and turned on again when the copper becomes too cool to work with. If a long soldering job is to be done the electric copper will facilitate matters very much owing to the fact that it can be continuously applied without reheating. In the case of ordinary coppers, it is necessary to reheat them occasionally to maintain a proper operating temperature.

Electric soldering coppers can be obtained for various purposes in various weights and shapes. A suitable shape for ordinary use is shown in Fig. 12.

AN ELECTRIC HEATER FOR SOLDERING COPPERS

The little device shown in Fig. 13 is a great convenience about the workshop. With it, the copper can be kept in a heated condition ready for use without employing torches of any sort. Furthermore, little trouble and expense are necessary to produce this electric heater.

The heating element is shown in Fig. 14. This is made up by winding 60 feet of No. 22 Climax resistance wire

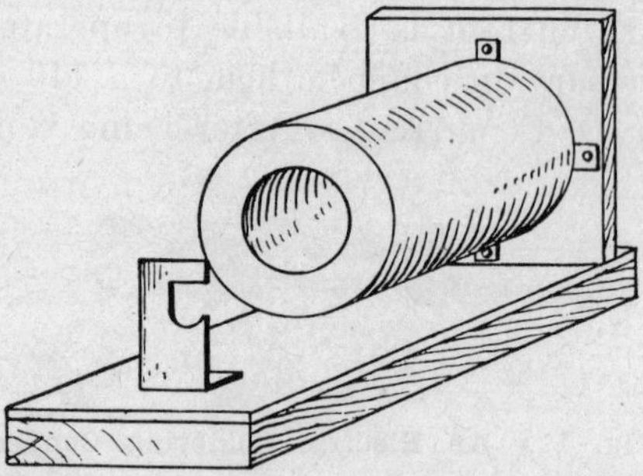

Fig. 13.—An Electric Device for Heating Soldering Coppers.

on a brass or steel tube 1½ inches in diameter and 3½ inches long. Before the wire is wound in place it will be necessary to cover the outside of the tube used with a couple of layers of very thin sheet mica. The first layer of resistance wire is wound over the mica. The resistance wire should be wound on so that the turns will not touch one another. This can be done by winding a string on the tube with the wire. After the wire has been put in place the spring can be taken off. If the wire has been wound on tightly it will stay in position. Two layers of mica are then placed over the first layer of resistance wire. The second layer of resistance wire is then placed over the first one with the mica interposed between the two. The second layer is wound in a direction opposite to the first one so that

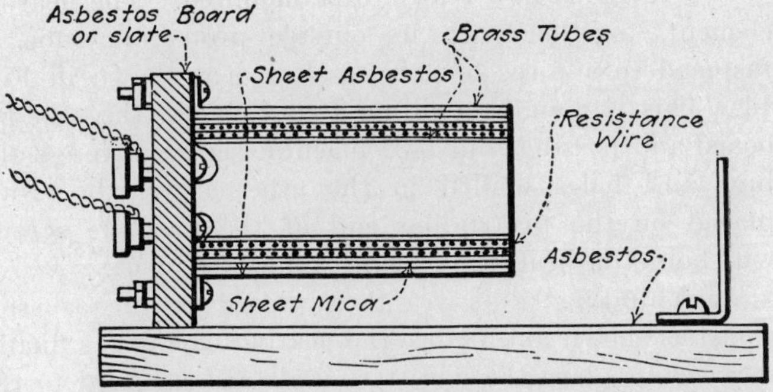

Fig. 14.—Cross Section of the Electric Heating Device, Showing Its Construction.

a non-inductive winding will be produced that can be used for either alternating or direct current. With the 60 feet of wire, 30 feet should be wound on each layer.

A second brass tube 3 inches in diameter and 3½

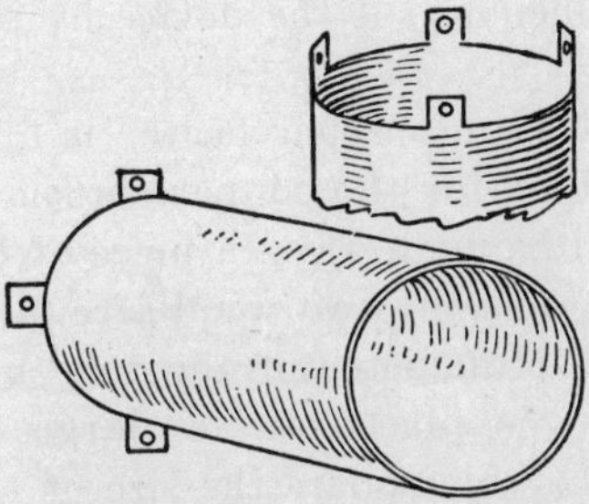

Fig. 15.—How the Cover of the Heating Element is Cut.

inches long is now obtained. This is cut so as to produce four lugs, as shown in Fig. 15. These are drilled out to accommodate a 6-32 machine screw.

The heating element is then wound with sheet asbestos

until it will just fit into the large tube. The fit should be as tight as possible without danger of disturbing the winding.

The device is now ready to be mounted. The heating element, together with its outside protecting tube, is fastened to a piece of asbestos board as shown in Fig. 14. This is done by drilling four holes in the asbestos board and passing four 6-32 machine screws through the lugs and holes drilled in the asbestos board. Nuts placed on the protruding end of the machine screws will hold the heating element rigidly in place.

Two binding posts are then placed in the asbestos board as shown and before the heating element is finally screwed in place the wire terminals are fastened to the binding posts. The asbestos board is then screwed to a base board of wood. The top of the base board should be covered with several thicknesses of sheet asbestos to prevent possible fire. Sheet asbestos is not only a refractory substance, but it also assists in thermal insulation. That is why it was used to wrap around the heating element. This provides a very noticeable increase in the efficiency of the device by preventing heat from radiating.

The little soldering copper heater is finished by placing a small holder for the copper, as shown in Fig. 13. This holder can be made from a piece of $\frac{1}{16}$ sheet brass. It is held to the base by two wood screws.

If the internal diameter of the tube upon which the heating element is placed is not large enough it can be increased to accommodate the size of the copper that the builder intends to heat. The same amount of resistance wire can be used with the device, even though the diameter of the tube upon which it is wound is larger.

AN ELECTRICALLY HEATED SOLDERING COPPER

With very little work the mechanic can convert an ordinary soldering copper into an electric type which will compare very favorably with the commercial devices. The first thing necessary to do is to wind two layers of thin sheet mica around the upper part of the copper to be converted. Over this is wound 10 feet of No. 26 Nichrome resistance wire. When this is wound in place with the turns properly spaced, another piece

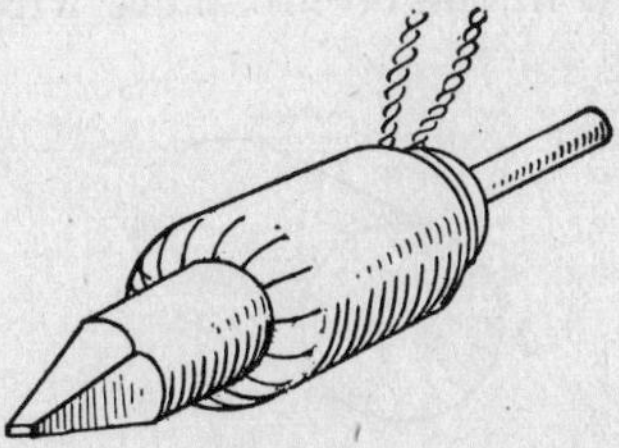

Fig. 16.—The Completed Electric Heating Soldering Copper.

of mica is placed over the winding. The winding is then continued and 10 more feet are wound on in the opposite direction to produce a non-conductive circuit, as described in connection with the electric heater in the foregoing paragraphs. Two more layers of sheet mica are wound over the second. When this is in place the heating element is covered with three layers of thin sheet asbestos. A small copper, brass or steel tube is then found with a diameter just large enough to slip over the complete winding. One end of the tube should be cut, as shown in Fig. 17. The purpose of cutting this tube will be seen by referring to Fig. 16, which shows the completed soldering copper. At the point where the terminals protrude a part of the tube is cut away to prevent a possible short circuit. The lamp cord is wound tightly around the terminals of the resistance wires. Solder cannot be used for this operation as the wires are maintained at too high a temperature in the immediate vicinity of the heating element. This would prevent the solder from staying in place.

This simple electric soldering copper can be used on any 110 volt lighting circuit on either D. C. or A. C. current. When it is placed in the circuit it will require by from five to ten minutes to bring it to a soldering heat.

If it is found that the soldering copper becomes too hot when placed in the circuit more wire can be wound

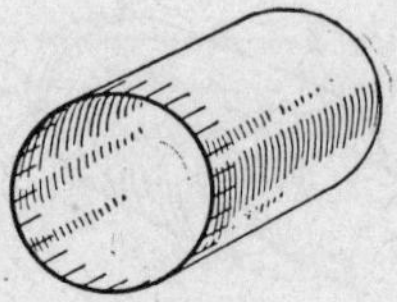

Fig. 17.—How the Tube is Cut to Fit Over the Heating Element.

around it. If the copper does not become hot enough, it will be necessary to take some of the wire off that has been already wound on.

AN ELECTRIC SOLDERING COPPER

The constructional details of an electric soldering copper that any amateur mechanic can make are given in Fig. 18.

The handle is of wood turned up to the dimensions given. This should be made from some close-grained wood such as cedar or spruce. After it is turned up on the lathe it is carefully papered to a smooth surface. A piece of iron gas pipe is then forced onto the open end of the handle as illustrated. If a piece of pipe cannot be found with the proper diameter a larger piece can be used after it is turned down upon the lathe. While the pipe should fit tightly, it should not be so large as to split the handle. The pipe should be forced for a distance of about 1½ inches back into the handle. About ¾ inch should be left protruding. A ¾ inch brass tube 4½ inches long is slipped over the protruding end of the iron pipe. A hole is drilled completely through the brass tube and iron pipe. This is tapped out to accommodate a 6-32 machine screw.

The heating element is now ready to be made. Twenty-

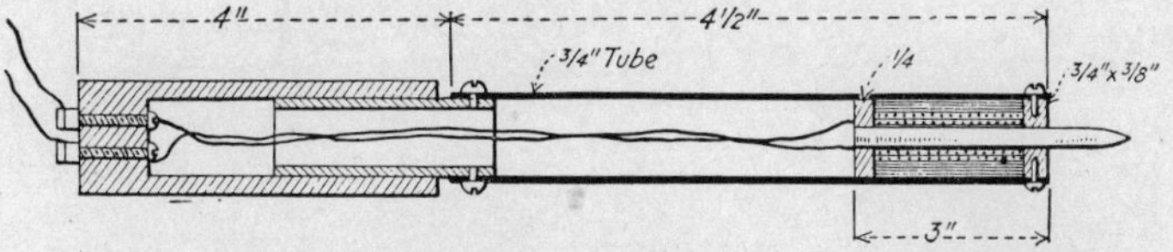

Fig. 18.—Constructional Details of a Homemade Electric Soldering Copper.

eight feet of No. 34 German silver resistance wire will be necessary to form the heating element. This is wound in place upon a copper or brass tube 3 inches long with an internal diameter of 5/16 inch. Thin mica is then wound over the tube. The first layer of resistance wire is then wound in place. The second layer is kept apart from the first one by interposing sheet mica as was done in making the previously described devices. Before winding the wire in place on the tube a small brass collar ⅜ inch thick and ¾ inch in diameter is turned up and forced over the outside of the tube upon which the heating element is to be wound. The opposite end of the tube is provided with a collar turned to shape from asbestos board or some similar refractive material.

After the resistance wire has been wound in place it is covered up with sheet asbestos. Just enough sheet asbestos is used to permit the heating element to slide into the larger brass tube freely. Two small holes are drilled with a No. 36 drill through the end of the brass tube into the brass collar, that was placed on the tube upon which the heating element is wound. These holes are then tapped out to accommodate a 6-32 brass machine screw. Before the device is finally assembled, two small holes are drilled in the end of the handle. 8-32 brass machine screws are placed in these as shown.

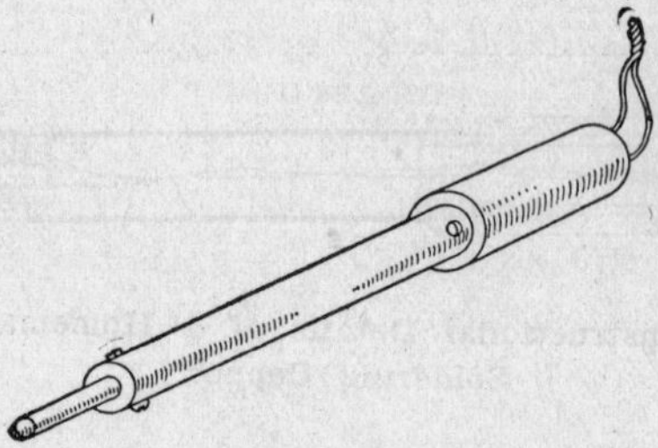

Fig. 19.—The Completed Soldering Copper.

Two insulated wires run from the brass screws to the terminals of the resistance wiring. The connection should be made by twisting the wires tightly around each other and pinching them together. It might be advisable to place some friction tape over the bare screws to prevent a shock when the device is being used.

A copper rod $5/16$ inch in diameter and $4\frac{1}{2}$ inches long is used to place inside the heating element. This is filed to a point in the lathe, as depicted in the sketch, Fig. 18. Eight holes are now drilled in the large tube. These act as ventilators and prevent the handle from slipping when the device is in use.

SOLDERING ELECTRIC WIRES

The right and wrong methods of twisting electric wires together before applying solder is shown in Fig. 20, A and B. After the wires have been twisted as shown

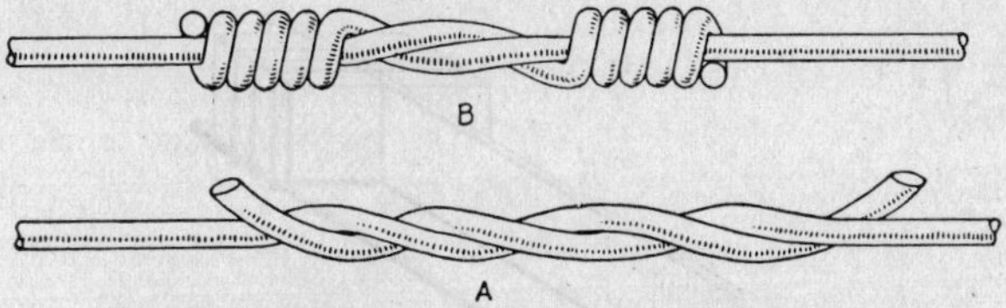

Fig. 20.—How Electric Wires Should be Soldered. A—Shows the Wrong Method. B—Shows the Right Method.

at B they are pinched together tightly with a pair of pliers to insure good contact. They should then be covered with a thin layer of soldering paste. The hot soldering copper should then be placed in contact with the joint and held there for a few seconds. The wire

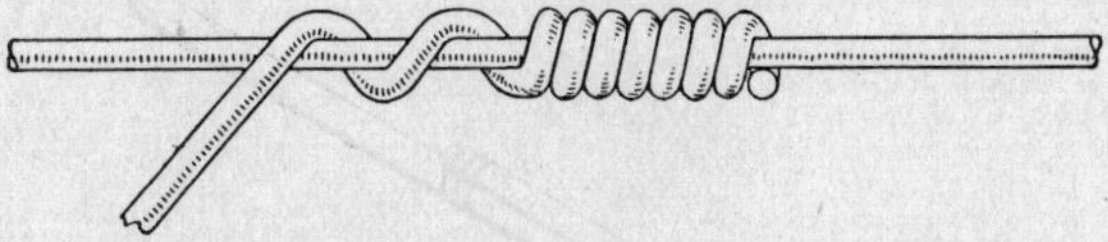

Fig. 21.—Method of Taking Off a Connection.

solder is then applied directly to the joint and it will flow nicely into crevices, forming a very good connection.

The method of taking a connection off a wire is shown in Fig. 21. This joint is soldered in the same way as the joint described in the preceding paragraph.

USE OF WIRE FOR HOLDING ARTICLES TO BE SOLDERED

Oftentimes wire can be conveniently employed for holding pieces together that are to be soldered. A notable example of the use of wire is shown in Fig.

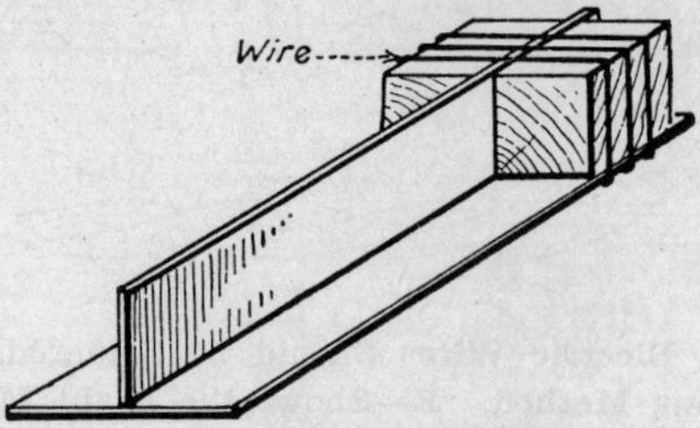

Fig. 22.—This Shows the Use of Wires in Holding Articles to be Soldered.

22. The one piece is to be soldered at exact right angles to another piece. Two small perfectly square blocks are pressed into service and one of these is placed on each side of the piece to be soldered. Wire is then

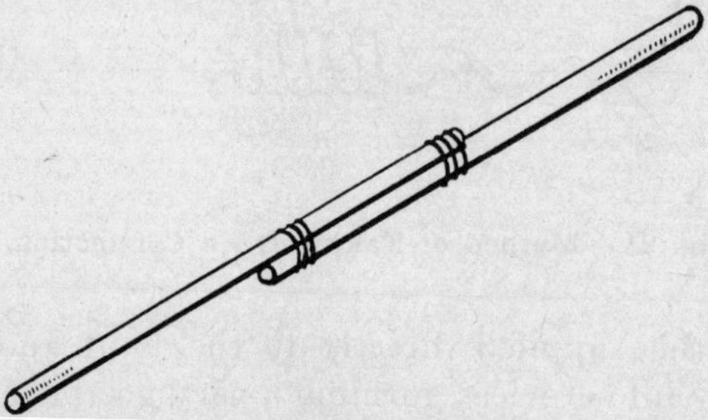

Fig. 23.—Another Use of Wire in Holding Articles Together.

wound around the blocks and metal pieces. The solder is then flowed into the two corners one-half of the length of the piece. After this is cool, the wire is taken off and the soldering is completed.

Another instance where wire can be used with convenience is shown in Fig. 23. Here two metal rods are to be soldered together. They are wired as shown and the wire should be drawn as tightly as possible, so that a good strong joint will be produced.

The mechanic will find that he can use wire to bind articles together in a great number of cases. When an ordinary soldering copper is used with solder wire it is very difficult to hold the pieces to be soldered and to manipulate the copper and solder at the same time.

SOLDERING LEAD

Lead cannot be soldered by the ordinary process, owing to the fact that the melting point of lead is nearly the same as the solder alloy. The electrical method described in the following lines will be found very suit-

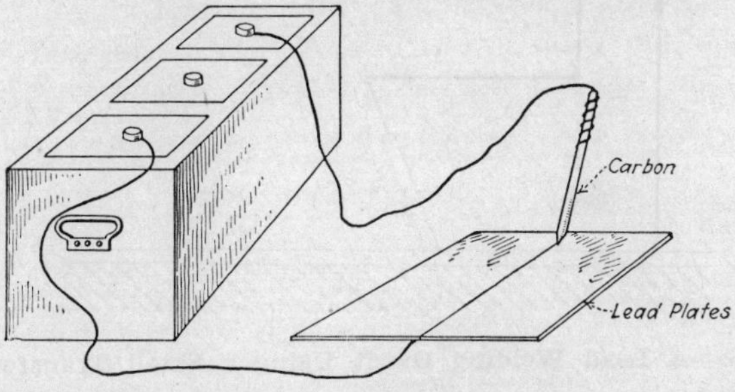

Fig. 24.—How Lead Can be Welded Together With a Storage Battery and a Small Carbon.

able in case two lead sheets are to be joined together, such as in the making of storage battery plates or the lining of wooden tanks with sheet lead, etc.

A small pencil of carbon about ¼ inch in diameter is connected to one terminal of a 60 ampere hour storage battery. The other terminal of the battery is connected to the lead plates or pieces to be soldered together. When the carbon pencil is brought in contact with the lead plates enough heat will be generated to melt the

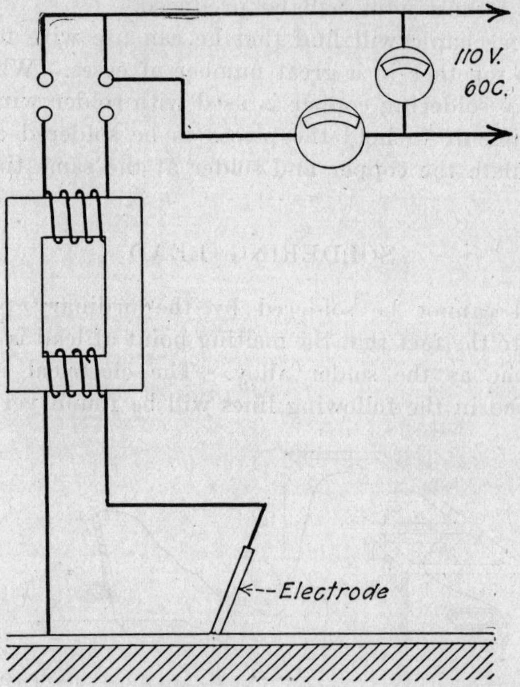

Fig. 25.—A Lead Welding Outfit Using a Small Transformer.

lead and form a metallic joint. The connections for the outfit are shown in Fig. 24.

This method has been found to work very satisfactorily. It is not necessary to use soldering flux as the hot carbon vapor produced by the spark acts as a reducing agent for the lead oxide which has been formed.

A more practical arrangement for soldering lead by the electrical method is shown in Fig. 25. Here a small welding transformer is pressed into service. It will be understood that the ordinary 110 volt lighting circuit could not be used for purposes of this nature, owing to the high potential that would tend to form a destructive arc. The potential used in this system should not be over 10 volts.

SWEATING

Another process, sometimes called "sweating," consists of tinning separately the parts to be united, after which they are placed in contact with each other and firmly held in position while being heated. After fusion of the solder, the object has its several pieces again brought as closely into contact as possible, either by tapping with a hammer, squeezing in a vise, by pressing together by hand, or by any means possible. Then, the work is left to cool, the superfluous solder removed, and a perfect solder joint is the result, provided the manipulations have all been properly carried out. The method last described is commonly used in the machine shop for uniting for use during turning or planing processes, the parts of a ring or bushing which must be in two or more pieces after completion. The several parts are fitted together, sweated in a continuous ring and then machined, after which they are heated, whereupon they fall apart as soon as the solder melts. The solder is removed, by wiping the surfaces with a piece of soft cloth and then taking off the balance of the solder which is in the shape of a thin film, with a scraper.

There are several special operations in soldering, where special work has to be done, and, in every instance, the specials are only adaptations of the ordinary process of soldering, to suit the particular work to be done. For instance: When a number of small or very small parts have to be soldered together in such a manner that the soldering of one would cause the others to become unsoldered, it is customary to hold each and every piece by means of clips and screws, or by means of clamps or weights. For some kinds of work, something very different may be required to hold all the pieces, and in some cases, it is necessary to put the parts in place one by one, and hold them there by means of calcined plaster, applied in the shape of cream. Each part is held in position until the plaster sets, after which each will stay in place until the solder can be applied.

For small work, which has to be thus held in place, the soldering copper is frequently too large to get into the small corners, between the several parts, and a good deal of trouble is frequently met with in getting the solder properly distributed. For work of this kind, the blowpipe is the most desirable tool. The blowtorch may be substituted for the blow-pipe if desired, but the former tool enables the heat to be localized better than with the torch. In either case, apply the heat until the solder flows readily, then with a short bit of copper wire, fastened to the end of an iron wire handle, the solder may readily be made to flow where it is needed.

If the piece of thick copper wire on the end of an iron wire handle is not at hand, a piece of copper wire may be used with good results to poke the solder around into the joints, but the trouble with the solid copper wire is that copper is a much better conductor of heat than the iron and the solid copper wire used as a small soldering bit, speedily becomes hot along its entire length and in so doing takes so much heat away from the business end of the tool that it will not continue to melt the solder.

The value of a soldering job frequently depends upon the use of a proper soldering fluid, and the stability of the soldered joint often depends largely thereupon. When acid is used there is little possibility that the joint will be permanent, unless means are taken for removing the excess of acid. However, this matter is much of a puzzle to some very advanced engineers, some acid soldered joints lasting apparently as long as those soldered without the use of acid, other acid soldered joints coming to pieces very quickly after their making. The electrical workmen have solved the question of the durability of the acid soldering joint by prohibiting its use entirely in electrical work.

METHODS OF SOLDERING VARIOUS METALS

Tin-Plate

When tin-plate is new and bright it is possible to solder it without the use of any flux. Rosin is the best flux for the purpose, however. If the tin-plate is old and tarnished it is necessary to use the chloride-of-zinc flux after the metal has been thoroughly cleaned by scraping or rubbing with emery-cloth until the surface is fresh and bright. After the soldering is finished the article should be carefully washed and dried to remove all traces of the zinc chloride and prevent the possible corrosion of the protecting layer of tin. Tin-plate is one of the easiest substances there is to solder, for the reasons that it is already "tinned" and is thin enough to absorb the heat quickly so that the solder will run very freely.

Galvanized Iron and Zinc

Galvanized iron is iron or low steel covered with a layer of zinc to protect it from the action of time and the elements. Both galvanized iron and sheet zinc are best soldered with a bit, using chloride of zinc as a flux. If the chloride-of-zinc solution is mixed with one-fourth its volume of hydrochloric acid the solder will oftentimes flow more readily, the action of the "free" acid being to dissolve the oxide from the surface of the metal and form more chloride of zinc. The work should be washed after the soldering is finished to remove all traces of the acid.

Lead

In making joints in lead it is sometimes desirable to employ a method of autogenous soldering called "lead-burning." The result is stronger and cleaner work.

Lead-burning consists in melting the metals and causing the parts to flow together and become joined without the aid of solder. It requires considerably more skill than any other form of brazing or soldering. A long step toward success may be taken by the proper arrangement of the work. It is usual to provide something which may serve as a mold or guide for the melted metal. For example, if two lead sheets are to be united by soldering, they are laid edge to edge on a sheet of some non-heat-conducting substance, such as brick or asbestos. The work in the immediate neighborhood of the joint is carefully scraped so as to remove all oxide or scale which would tend to bind the melted lead and prevent it from flowing freely. The metal at the seam is melted by a very hot bit or the flame from a blow-pipe so that there is a uniform flow of lead across the seam. It is sometimes necessary to add more lead to the seam by melting a strip held in the hand.

A flame of some sort is the most satisfactory source of heat for the average lead-burning job, because not only is the heat more uniform, but also more intense, and the lead melts at the desired point before the surrounding metal becomes sufficiently hot to soften.

There are several types of blow-pipe for this purpose

on the market. Some employ an alcohol flame, while others make use of mixed hydrogen and air. The flame is usually small, sharp-pointed, and very intense.

Lead-burning is absolutely necessary, and is insisted upon in certain classes of work, for instance, in lining tanks with lead for chemical solutions, or for joining the grids and lugs of storage batteries. The presence of even the smallest amount of solder containing a foreign metal such as tin causes a local action to be set up, and the chemicals slowly attack the seam at that point. Most of the large manufacturers of storage cells also make a pair of lead-burning tongs which are of the proper size and shape to fit around the lug of their cells. These can be purchased for repair work, and provide a snug mold, so that all that is necessary is to flow the lead with the flame and allow it to cool.

Lead Pipes

Where conditions preclude the possibility of cutting a thread, it is sometimes necessary to join an iron pipe to a lead pipe by soldering. In such a case the end of the iron pipe is first carefully scraped and then tinned with the aid of some "sodium solder." Sodium solder is not a necessity, but makes possible a much quicker and easier operation.

Sodium solder is prepared by dropping small pieces of metallic sodium into a ladle full of melted solder. The sodium will melt and become absorbed by the solder with a slight puff of smoke. Most of the prepared solders which are on the market, and readily adhere to metals without the aid of a flux, contain a certain percentage of sodium. Sodium solder can be used very successfully for tinning, but joints made wholly with it tend to corrode.

After the iron pipe has been scraped and tinned it is

an easy operation to solder the lead pipe to it, using tallow as a flux.

When brass fittings are soldered to an iron pipe it is the most desirable to use a blow-pipe if possible, because it performs in this case a more satisfactory job than the bit or copper.

The end of the lead pipe is expanded slightly by driving in a turnpin, and then carefully scraped inside. The clean portion is covered with rosin or tallow. The end of the brass fitting is cleaned and tinned, and, after being driven into place, a blow-pipe flame directed upon the joint. More heat should be given to the brass than to the lead. The heat should be just sufficient so that

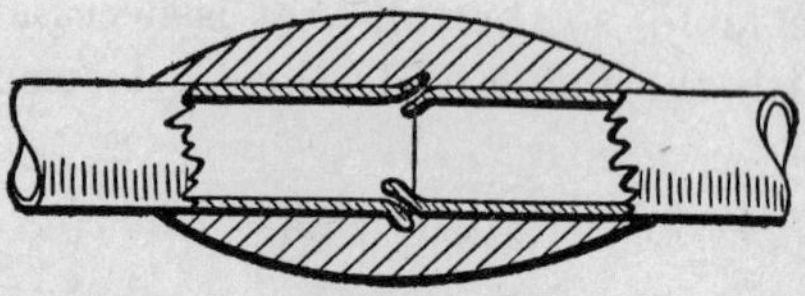

Fig. 26.—Method of Making a Wiped Joint.

the solder will flow freely into the joint, but the pipe itself will not melt. At the instant of complete fusion a quick wipe around the joint with tallow will finish and smooth the job. (See Fig. 26.)

The "wiped" joint so commonly employed by plumbers in fitting lead pipes where water pressure must be withstood is an operation requiring skill and practice. The joint is prepared for wiping by first expanding one side so that the other will fit snugly into it. After the metal has been thoroughly cleaned it is smeared with tallow and put in position. "Plumber's" solder is splashed or poured over the joint until the metal adjacent is heated almost to the melting-point, and the solder flows over joint and amalgamates with it. The running solder is immediately caught in a wiping-pad held underneath the joint and worked around the joint. As the solder becomes cooler it changes to a crystalline and granular state, and is "mushy." A wiping-pad is composed of several thicknesses of heavy duck saturated with tallow.

The portions of the work beyond the joint, where it is desirable that no solder should appear, are usually painted with a mixture of size and lampblack.

The same principle is employed in "wiping" joints of seams in sheet lead. The boundaries are painted or "soiled" with lampblack so that the solder will not adhere to those portions. The area to be covered is scraped clean and rubbed with tallow. The solder may be poured on or melted on from a stick heated with a torch or blow-pipe. Perhaps the last method is the best. As soon as the work has been heated sufficiently to become plastic, the seam is shaped with a wiping-pad or with a special-shaped piece of wood of asbestos.

Aluminum

Aluminum is the most difficult metal to solder. It was a long time before any satisfactory solder for joining this metal could be discovered, and for a time it was thought impossible.

The aluminum solders now on the market consist of "straight" aluminum solders and "combination" solders. The "straight" solder is used for soldering aluminum to aluminum and the "combination" solder for soldering any metals, copper, brass, etc. (except cast iron) to aluminum.

Parts to be soldered are heated with the flame of a blow-pipe until the solder begins to melt upon them. Then brush the melted solder well into the metal. If any dross forms on the surface, brush or scrape it off well. Each part should be tinned separately and then pressed together and soldered.

To use combination solder, "tin" the metal to be soldered to the aluminum with some of the "combination," using any good flux for the purpose. After cooling remove all traces of the flux and prepare the aluminum piece with some of the "straight" solder, as described above. The parts may then be joined with some of the "straight" solder without the use of any flux.

The "combination" solder is merely for tinning purposes, while the "straight" solder is for the actual soldering.

PART II

HARD SOLDERING AND BRAZING

Where strong joints capable of withstanding considerable mechanical strain are necessary, some method other than soft soldering must be employed. For this purpose alloys known as hard solders have been brought into use. Hard soldering alloys are made up of metals melting at a higher temperature than the ordinary lead-tin mixture. Owing to the higher temperature necessary for use with hard soldering, it is a much more difficult process than that of soft soldering. The mechanic will find it necessary to practice considerably before he will be able to handle this process with any degree of success.

Hard soldering will be considered first. In fact, the process of hard soldering and brazing are very closely allied. Hard soldering employing a silver-brass alloy is sometimes termed silver soldering. Brazing is really the process of uniting two metallic surfaces together with brazing spelter.

Silver solder consists of pure metallic silver and brazing spelter mixed together in various proportions. The percentage of the metals in the composition determines the temperature at which the alloy melts. In this way hard soldering alloys can be produced with a melting point anywhere from 700° Fahr., to 2,000° Fahr. It will be understood that the higher the melting point of the solder used the stronger the resulting joint will be and vice versa. In this way it will be seen that the melting point is a direct indication of the mechanical strength of the solder. Hard solders with a melting point in the neighborhood of 700° produces a joint with little more strength than the ordinary mixture of lead and tin. On the other hand, a hard solder melting at 2,000° Fahr. produces a joint extremely strong and will stand great abuse before finally breaking.

One thing must be kept in mind. A hard solder should not be used with a melting point that is too close to the melting point of the metal pieces that it is used to join together. A good silver solder to be used in connection with copper consists of two parts silver to one part brass in the form of brazing spelter. A good mixture for work with brass is made up of seven parts of silver and two parts of brazing spelter. Silver solder already prepared in sheet form can be purchased from large jeweler's supply houses. In ordering silver solder the mechanic should specify what particular grade he desires as it is made with various melting points. The mechanic can also melt up his own ingredients and roll it out into a sheet if he so desires. This is probably the most suitable form to use it in as it does not require very great time to melt.

The various tools and materials used in the process of hard soldering are shown in Fig. 27. The outfit is not elaborate but it will nevertheless enable the mechanic to accomplish all ordinary jobs of silver soldering where an alloy is used that does not have an extremely high melting point.

The heat used in silver soldering must be very intense and for large pieces of work it is necessary to employ a big flame. The worker will realize that a small flame, no matter how hot, does not produce a great volume of heat. The ordinary gasoline blow torch produces a hot flame for this work and it has sufficient heat to melt the alloys with a lower melting point.

The use of the various tools and materials in carrying on the process of silver soldering will now be explained.

An acid pickle is made by mixing one part of sulphuric acid with twenty parts of water. This pickle is used to thoroughly clean the article after it has been silver soldered. The borax which is used as a flux is baked hard on the surface of the metal when the process is finished and the sulphuric acid pickle is the only practical and efficient method for quickly removing it. This pickle can also be used in cleaning up an article before

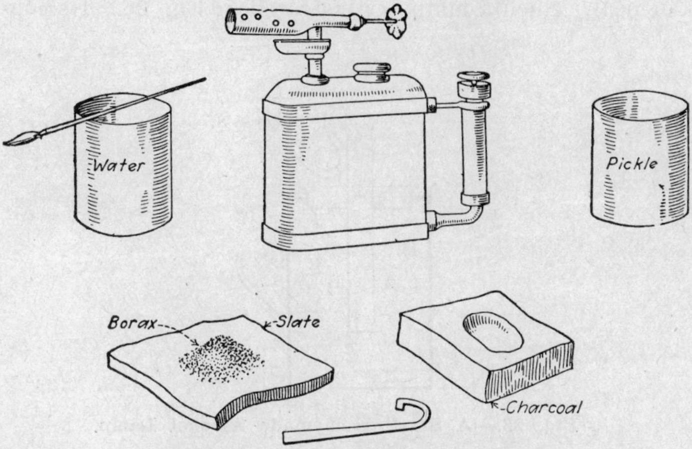

Fig. 27.—A Hard Soldering Outfit.

the solder is applied. Of course, the ordinary process of scraping and polishing must also be resorted to to bring the metallic surface to a proper soldering condition. A charcoal block is used to place the work upon while the soldering is being done. The object of this block is to return the heat to the work which helps greatly in speeding up an operation. In some cases, however, it will be found that it is quite impractical to use the block, even though it would help the work. On a large piece of work where it would be impossible to such a block the work can be placed in a sheet iron receptacle filled almost to the top with charcoal. The borax is used as a soldering flux and is employed in the same way as rosin is used in ordinary soft soldering. The borax is first moistened and rubbed on a piece of slate. This process produces a thick paste of borax and water. This paste is then painted on the work where the soldering is to be done.

Where a very small piece of work is to be silver soldered a small alcohol lamp can be used as a source of heat. Such a lamp is illustrated in Fig. 28. Its construction is extremely simple and the mechanic would have little trouble in assembling it. As before stated, this torch will only furnish sufficient heat for very small jobs, owing to the small volume of heat produced by the flame. It will be understood that the metal tube must be blown into at the cold end.

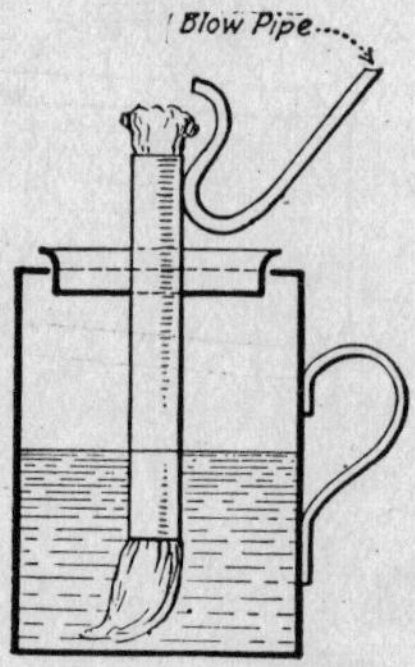

Fig. 28.—A Small Homemade Alcohol Lamp.

For the sake of convenience, it will be assumed that the two pieces of metal shown in Fig. 29 are to be silver soldered. First it will be necessary to wire the pieces together so that the solder can be applied conveniently. Unless the pieces are wired in this way it will be difficult to bring them both to the same temperature. It would also be difficult to hold them while the soldering was being done. It is assumed that the pieces were thoroughly cleansed before being wired together.

After the pieces are wired together, borax paste should be smeared around the joints. The article is now placed in the flame and is left there until it becomes

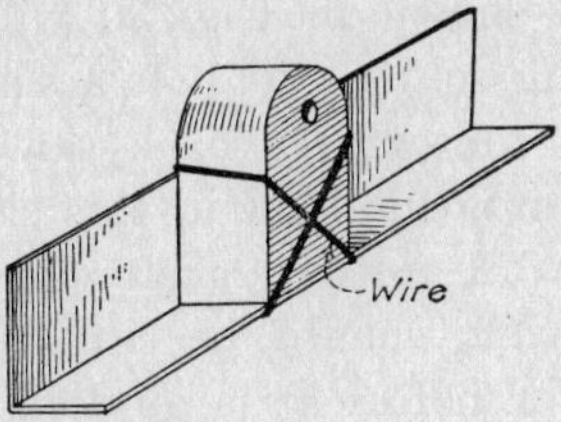

Fig. 29.—Pieces to be Silver Soldered.

red hot. If the borax drops off during the heating process it should immediately be replaced with a piece of wire. This is done by placing some of the borax paste on the end of the wire and smearing it on the work in the flame. When the piece has become red hot, tiny pellets of silver solder should be put in place along

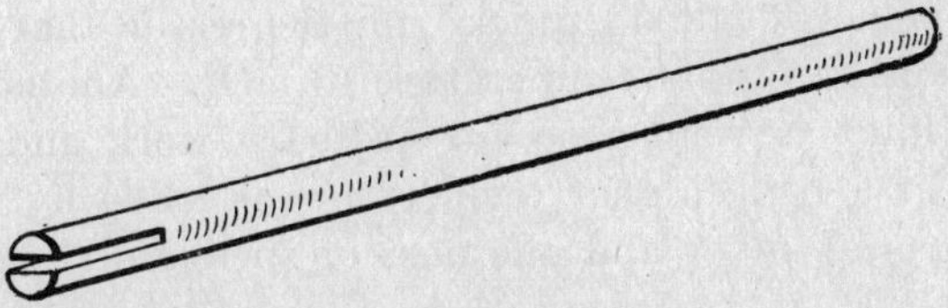

Fig. 30.—A Small Tool Used in Holding Silver Solder.

the crevices, as in Fig. 29. This can be done with a little pair of iron tongs. The piece is then left in the flame until the silver solder melts and runs into the joint. The piece is then turned around and the opposite corner is treated in the same manner. The two bottom corners are then soldered in the same way. The pieces should be taken from the flame and allowed to cool gradually before it is finally dipped in the acid pickle to remove the superfluous borax which will be baked to a crust on the surface. After removing the work in the pickle, it should be thoroughly rinsed off in clean water to further remove traces of the acid, which, if left, would tend to pit the surface of the metal.

The little tool shown in Fig. 30 will be found very useful in applying solder. This is a small piece of steel rod split at the end with a hack saw as shown. The sheet silver solder is inserted in this slit and the silver is held to the work until it melts and runs in place. The work, however, should be heated to the melting point of the silver before it is applied.

A good silver solder for general purposes of thin brass sheet can be made by mixing twelve parts of silver and one part of brass. This particular alloy has a comparatively low melting point and for this reason it can be used successfully with thin sheet brass. If a solder with too high a melting point was used to join thin sheet brass, it would be found that the brass would soften to such a point that it would be impossible to accomplish the work. A silver solder made according to the above proportions is called ''quick'' for the reason that it does not require a great temperature to melt. Another mixture which is very good for ordinary work and which melts at a little higher temperature is made by mixing six parts of silver and one part of spelter or brass.

PART III

BRAZING

Brazing is generally regarded as a process by means of which metallic articles are joined by an alloy made up of copper and zinc. Brazing spelter is available in the granular state, so that it will melt at the proper temperature. Of course, it is understood that the copper zinc alloy really produces a certain grade of brass. After the process of brazing is described more attention will be given to various alloys of copper and zinc, together with information regarding the control of the melting point by changing the percentages of the metals.

Perfectly clean surfaces must be produced in the process of brazing just as in hard and soft soldering. The worker with soldering processes should always remember to make the surface scrupulously clean. The parts to be brazed should be wired together as tightly as possible. The same is true here as is true with hard soldering and soft soldering alloys. Many mechanics believe that the surfaces should not be brought in contact too closely owing to the fact that the molten spelter will not find its way into the joint. This is quite contrary to the actual condition. Brazing spelter in a molten state will find its way into crevices of very small width. The resulting joint will be much stronger than a joint would be with the brazed surfaces a greater distance apart.

Owing to the high melting point of the various brazing alloys the workman must be careful not to attempt to solder metals with a lower melting point than the particular alloy used. This would result in the destruction of the pieces to be soldered.

Borax is the flux employed for all brazing processes. In commercial work which is carried on on a large scale, boric acid finds greater use than ordinary borax, owing to the fact that it is cheaper and can be purchased in granular form.

As before mentioned, brazing spelter consists of parts of copper and zinc. One of the most widely used combinations consists of fifty parts brass and fifty parts

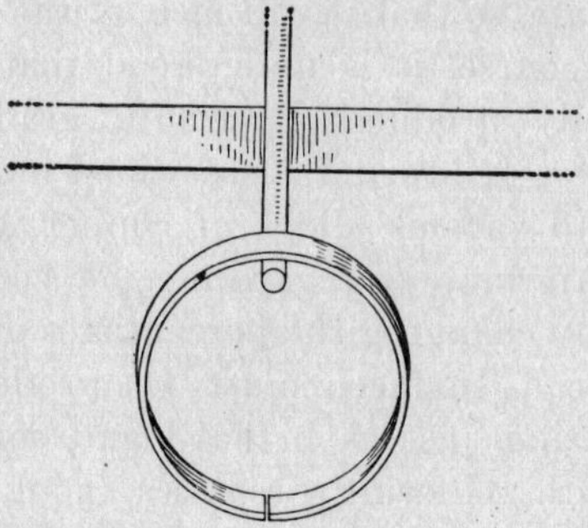

Fig. 31.—How a Bronze Ring Should be Held in Soldering.

zinc. For different operations it is necessary to use a harder or softer alloy and this will make necessary alloys of different composition.

In commercial brazing it is often profitable to mix the spelter with the proper proportion of boric acid. The "proper proportion" must be found by experiment. The mixture is placed over the parts to be brazed and sufficient heat is directed upon the spelter to cause it to melt and "run." As soon as the spelter runs the article is removed from the fire and allowed to cool. The superfluous flux can be removed by a scraper or with a wire brush. The solution used in connection with silver solder can also be employed effectively in bringing about the removal of the spelter.

The manner of applying the spelter and borax depends upon the job to be done. If a plain ring is to be brazed it should be hung from the end of a rod of iron so that the part to be brazed is downward, as illustrated in Fig. 31. A little borax and spelter are then placed on the joint to be soldered and a hot flame is afterward directed upon it until the spelter fuses and is deposited upon the joint. In applying the heat to the ring it is best to do so slowly as the temperature should not be carried too high in too short a time. Also the metal on each side of the joint should be heated to the same temperature, otherwise an insecure and poorly made joint will result.

The source of heat is a very important consideration. The flame used should be as free from soot and solid matter as possible. An unclean flame will be sure to deposit foreign matter upon the surfaces to be brazed, which will in turn prevent the spelter from adhering. A gasoline torch can be used very successfully on small jobs. However, for large work the ordinary gasoline blow torch flame does not have a sufficient volume of heat and is therefore quite useless. A splendid brazing outfit for all work of ordinary size is described in the last part of this book. The torch of this outfit delivers a flame several times larger than the flame of the ordinary gasoline torch.

By the aid of a bellows, brazing can often be accomplished in a clean coal fire or charcoal fire. A fire made of coke produces an immense amount of heat and also a nice clean fire. If this is used in a smith's forge a very good source of heat for brazing purposes is available.

If brazing is to be done in a smith's forge the bellows or blower should be operated until all the visible gaseous matter has passed off and a clean, hot fire results. With a poker a crater is made large enough in the center of the fire to hold the work. The work is then

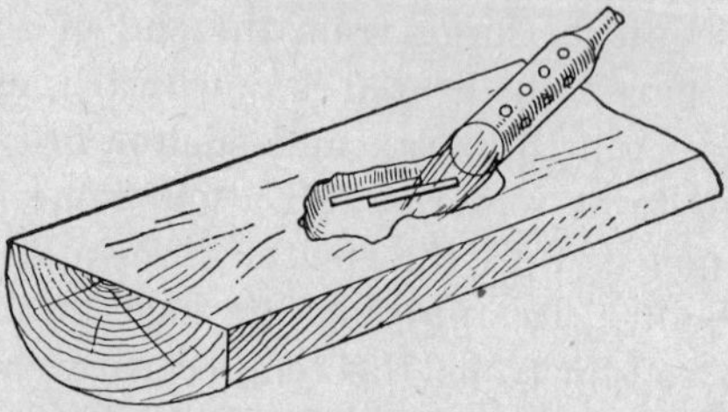

Fig. 32.—Pieces Placed in a Small Charcoal Block for Silver Soldering.

lowered into this crater and the operation of the blower is continued until the work has been brought to the proper temperature. It is understood, of course, that the borax and brazing spelter have been put in place

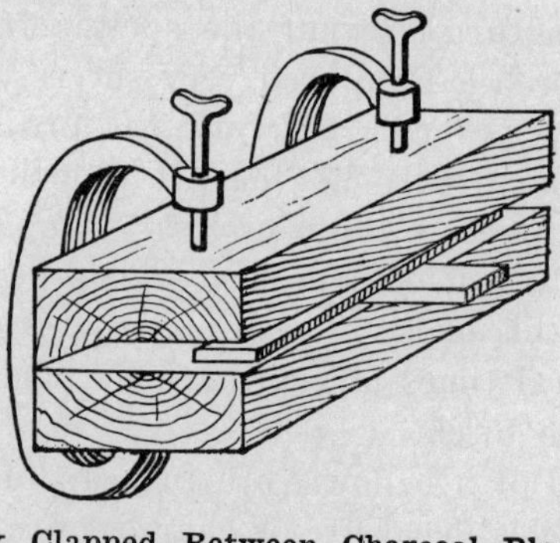

Fig. 33.—Work Clapped Between Charcoal Blocks for Silver Soldering.

on the joint before the work is lowered into the crater. The workman should carefully watch the heated piece and with a small iron wire shift the position of the borax and spelter should they become disturbed in any way. When spelter starts to run it may become rather unruly and if it is not carefully watched it will probably flow to the wrong place or off the work.

In case of brazing with a small gasoline flame the pieces to be joined can be placed on a small charcoal block, as illustrated in Fig. 32. Another very convenient and practical method of holding a piece of work to be brazed is shown in Fig. 33. Here the job is clamped upward between two charcoal blocks. These blocks are used to thermally insulate the work so that the proper temperature will be arrived at quickly.

BRAZING BY IMMERSION

A process is used for commercial purposes which is more rapid than the ordinary method of brazing and has become known as the immersion system. With this system the brazing spelter is kept in a molten condition

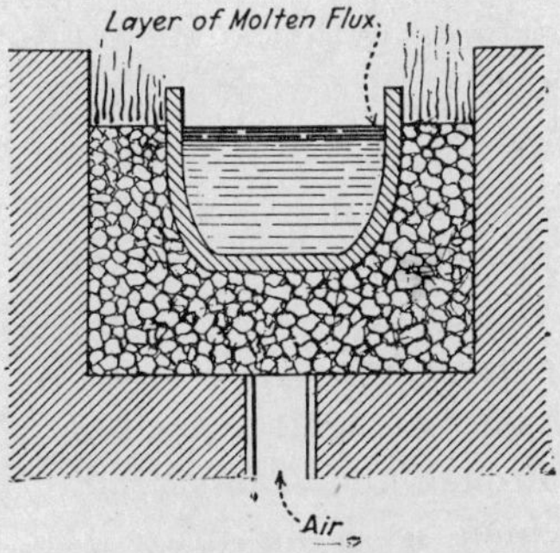

Fig. 34.—A Crucible for Brazing by Immersion.

in a large iron receptacle. A layer of molten flux is placed on top of the spelter. Thus, in introducing an article to be brazed it must first pass through the molten layer of flux before it reaches the spelter. This is shown clearly in Fig. 34. In the case shown in Fig. 34, the receptacle containing the flux is imbedded in a fire of hot coal with an artificial draft forced up from the bottom. A more suitable method is the employment of a gas-heated furnace by means of which a more uniform heat can be maintained and the necessity of replenishing the fuel is avoided.

When work is introduced into a molten bath of spelter in the manner described here, the spelter finds its way into every crack and crevice. Capillary attraction causes all the spaces to be filled.

In brazing by this method it is often necessary to prevent certain parts of the work from being covered with the molten spelter. This can be very easily done by smearing the parts not to be covered with graphite paste. This paste will so affect the surfaces that no spelter will adhere to them.

The choice of fluxes used in the immersion method of brazing is somewhat optional. Ordinary borax seems to be as good as any other mixture and it should be melted on the surfaces with the spelter to form a layer from three-quarters to three inches deep.

Another method of brazing is known as the "Pich Process of Brazing Cast Iron." Cast iron can be brazed by the methods described above, but it is very delicate business, as the iron melts, or at least softens so it will break under the least strain, at a temperature pretty near that at which the spelter melts, so that it is almost impossible to melt the brass without "burning" up the cast iron which is being brazed.

By the Pich method, the surfaces to be brazed are first brushed over with a varnish made of oxide of copper mixed with any liquid which will allow of the copper being spread with a brush, and which will afterwards hold the oxide when dry.

After this application, the brazing is carried out the same as in the ordinary method. The brass or spelter is placed in position, and the flux applied, then a gas-air flame is applied as in ordinary brazing. It is assumed that the metallic oxide acts as a reducer on the surface of the cast iron to be brazed. Without the oxide, the carbon above noted acts much like the graphite used in the dipping method of brazing as an anti-flux as described in the description of the dipping process of brazing. It is claimed that the metallic oxide is reduced, removing during the process of reduction the carbon on the surface of the metal, and, it is claimed, often penetrating for a distance of three or four inches into the metal itself, thereby making the cast iron stronger, at and near the joint, than it was before. Of this matter, the writer has no personal knowledge. The process has been described very fully in a paper by Wilfred Lewis, read before the American Society of Mechanical Engineers.

It is claimed that both the joint and the casting itself is made from 5% to 10% stronger by the treatment with oxide. This, if correct, is a pointer of value to the iron worker, aside from in the process of brazing, for, if in certain cases, cast iron can have its strength increased 10%, it will be of inestimable value to the designing engineer to know thereof.

BRAZING ALLOYS

As before mentioned, brazing alloys can be made with melting points varying between 1980° and 786° Fahr. A splendid table showing this increase and decrease of melting points according to the different percentages of copper and zinc is given on the next page.

ALLOYS OF COPPER AND ZINC

	Percentages in the solder										
COPPER:											
	100	90	80	70	60	50	40	30	20	10	0
ZINC:											
	0	10	20	30	40	50	60	70	80	90	100
MELTING POINT, FAHR.:											
	1983	1904	1823	1706	1652	1616	1508	1436	1292	1076	786

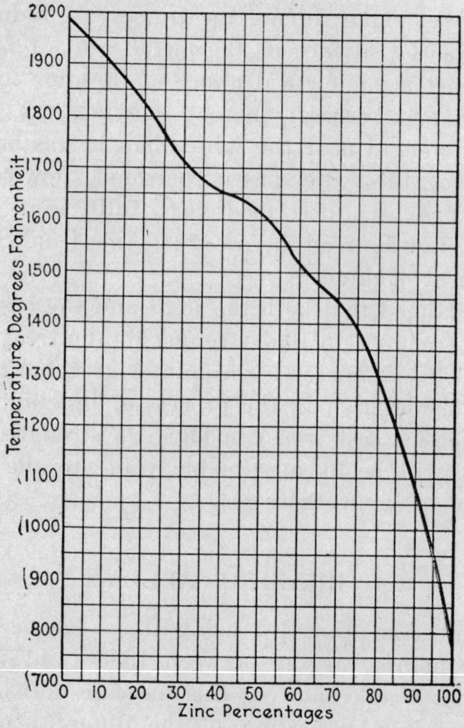

Fig. 35.—The Melting Point of Copper-Zinc Alloys of Various Percentages.

A curve illustrating the various melting points for different percentages of zinc is given in Fig. 35. If the worker wishes to know what mixture of copper and zinc will produce a low melting point of 1300° Fahr., he could lay a ruler parallel to the horizontal line with its top at the 1300 level. The ruler will then intersect the curve at a point to the right of the main vertical line on the left. This will give a reading of 78% zinc. Thus an alloy consisting of 22% copper and 78% zinc must be the result. This is assuming, however, that the copper and zinc used are absolutely pure. If the metals are purchased at a chemical supply house it is possible to produce an alloy with a melting point that will correspond very closely to that given on the table or curve.

The ordinary brazing spelter used for miscellaneous purposes is a half and half mixture of copper and zinc. It melts at a temperature of 1660° Fahr. and produces a good strong joint.

Coppersmiths' spelter is produced by mixing together three parts of copper and one part of zinc. According to the curve in Fig. 35 the melting point of this particular alloy will be in the neighborhood of 1762° Fahr.

What is known as gray solder is a popular brand and is very easily melted. It is produced according to the following formula:

Copper	44 parts
Zinc	50 parts
Tin	4 parts
Lead	2 parts

By changing the percentages of the constituents, brazing alloys may be produced with a wide variation of melting points. Another very good gray solder consists of the following ingredients:

Half-and-half soft solder	4 parts
Half-and-half hard solder	88 parts
Zinc	6 parts
Tin	2 parts

What is known as white solder is made as follows:

A 2:1 brass (copper, 2 parts, zinc, 1 part)—a medium strong coppersmiths' spelter	82 parts
A 1:3 lead-tin solder (lead, 1 part, tin, 3 parts)—a pretty soft solder	18 parts

A very good alloy known as white solder may be produced from spelter by adding tin as indicated.

Coppersmiths' spelter	85 parts
Tin	15 parts

Another very useful variety of half white solder that has found wide application is composed of:

Half-and-half hard spelter	92 parts
Lead-tin solder (lead, one part, tin, three parts)	8 parts

When mixing brazing alloys using lead tin composition, it must be remembered that the brass, which has a higher melting point, should be melted first. When the softer alloy is added the mixture should be stirred up in order to produce a perfectly homogeneous alloy. Brazing spelters which include nickel, gold and silver are often used. Nickel may be added to a brass spelter and in this way very hard and strong solder is produced. Brass nickel spelters are used many times in joining surfaces of German silver and steel. A good German silver spelter can be made up according to the following formula:

Copper	35 parts
Zinc	55.5 parts
Nickel	9.5 parts

Another very good solder for the same purpose is produced by alloying

Copper	38 parts
Zinc	50 parts
Nickel	12 parts

Still another German silver solder is made by mixing

Copper	36 parts
Zinc	56 parts
Nickel	8 parts

PART IV

HEATING DEVICES

The greatest trouble that the mechanic has in hard soldering is that of producing sufficient heat to carry on the operation. In this part of the book heating devices and methods will be described.

As has been stated before, the ordinary gasoline blow torch can be used successfully on jobs that do not require a great volume of heat. For the benefit of those

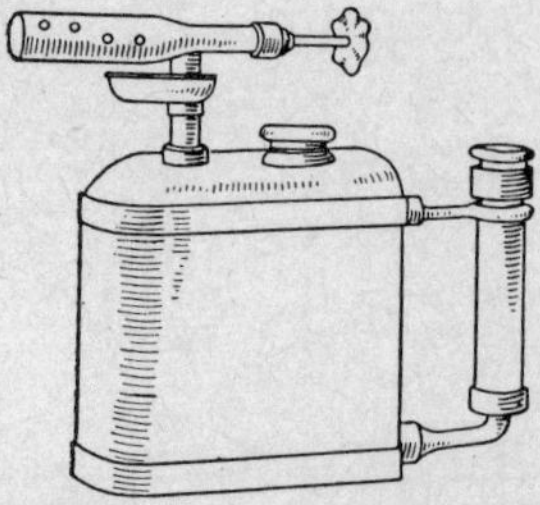

Fig. 36.—A Gasoline Blow-Torch.

who do not know how to operate one of these torches, a few words will be given to this subject. An ordinary gasoline blow torch is shown in Fig. 36. These torches are all provided with a small hand-operated air pump so that the air pressure can be produced within the tank containing the gasoline. The tank should, therefore, be filled about $3/4$ full to leave sufficient room in which to produce an air pressure of a few pounds. This air pressure forces a thin stream or spray of gasoline out of the needle hole when the valve is open. A small trough is placed under the burner and before the torch can be started it will be necessary to run a little gasoline into this trough by opening the valve. This gasoline is then ignited and allowed to burn away. Just before the flame disappears, the valve is opened again and by this time the torch will be sufficiently hot to vaporize the gasoline and produce an extremely

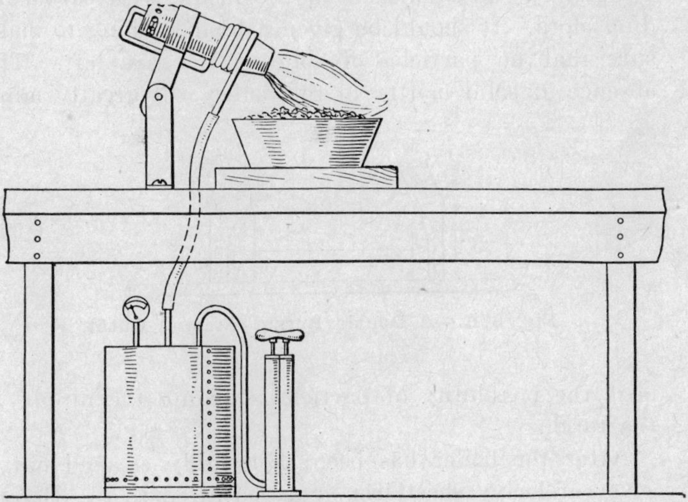

Fig. 37.—The Complete Blow-Torch Outfit.

hot flame. The air pressure within the tank must be replenished from time to time.

A larger gasoline burner for brazing can be made by the home mechanic if he will follow the directions outlined in the following paragraphs. The torch in question is shown completed in Fig. 37. This outfit will produce an extremely hot flame with sufficient volume to accomplish most of the ordinary jobs of hard soldering or brazing.

The tank for the outfit is made from the end of an old hot-water boiler. Such a boiler can be obtained from a local plumber for what it is worth as junk in most cases. The portion cut off is 24 inches long and the cutting can be done with a large hacksaw. A mark should be made around the boiler with a heavy pencil at the cutting point before the cut is attempted. This procedure will insure a straight cut. When the boiler is cut the portion of it to be used should be thoroughly scraped out and polished on the inside with carborundum cloth. It should be given a final washing to make sure that no particles of solid matter are left. The absence of solid matter in the boiler will greatly minimize the possibility of particles clogging the nipple of the torch.

Fig. 37A.—A Double Burner Brazing Outfit.

After the boiler has been thoroughly cleaned out a cover of heavy sheet iron must be fitted to it. Before placing this cover on the end, an internal metal ring is riveted in place, as illustrated in Fig. 38. The rivets that hold this ring in place should be covered with solder on the outside to prevent a possible leakage of air. This metal ring forms a seat for the cover. The cover should be cut as true as possible and it should make a tight fit into the top of the tank. Before finally placing the cover in position the holes for the air pipe, feed pipe and pressure gauge should be drilled. These holes should not be drilled after the cover is in place, as part of the metal chips from the drilling will fall inside the tank.

It will be necessary to braze the cover to the tank. In most cases it will be necessary for the mechanic to take this job to a local welding or brazing shop. After the brazing is done the tank should be tested to see that it has no serious air leaks. A good way of doing this is to put the pressure gauge in place, plug the feed pipe hole up and place the air pipe in position. Owing to the thickness of the cover it will be necessary to place a small collar over each hole so that sufficient threads can be cut to hold the piece without danger of air leakage. These collars should be brazed or soldered to the top of the tank. If the tank is placed under water and

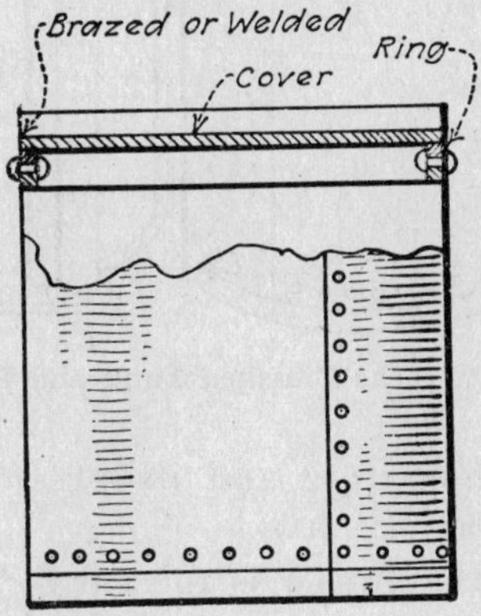

Fig. 38.—How the Top of the Tank is Held in Place.

air pumped into it any leakage will be noticed by bubbles rising to the surface. If no bubbles rise to the surface, the tank may be regarded as air tight and attention directed to the construction of the burner. The finished tank and pump are shown in Fig. 39. The burner and its details are shown in Fig. 40. Before describing the actual construction it may be well for the reader to thoroughly understand its operation. The coil which is wound around the burner is called the vaporizing coil and the gasoline enters the pipe at A from the tank. It must circulate through the hot vaporizing coil before it reaches the nipple B, where it is finally ejected into the burner for combustion. When the gasoline enters the vaporizing coils it is heated to a

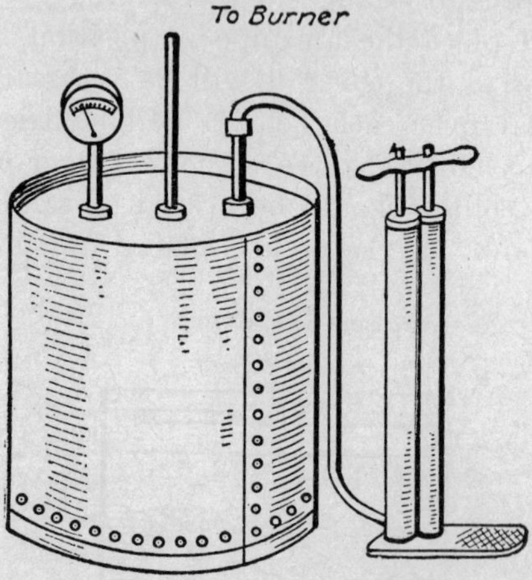

Fig. 39.—The Finished Tank and Pump.

rather high temperature and its pressure thereby increases considerably.

The body of the burner is formed by a sheet of iron bent into a cylinder which measures 5 inches long, 2

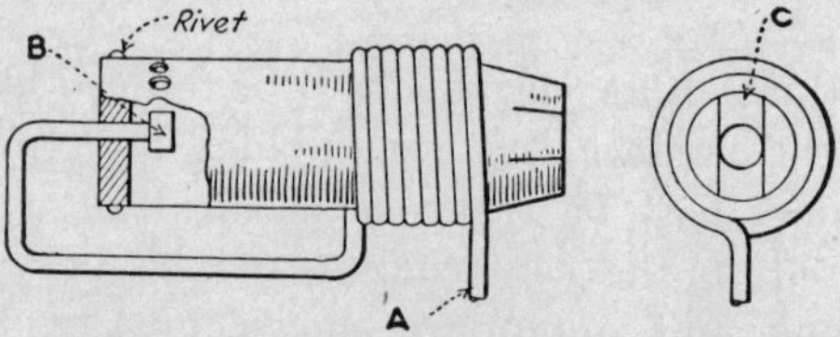

Fig. 40.—Details of the Burner.

inches in diameter. One end of the iron cylinder is cut in about ½ inch so that it can be bent as shown in Fig. 40. This is to produce a small aperture at the end where the flame is ejected. About ¾ inch in back of the burner a number of holes are drilled to permit free entrance of air, so that sufficient oxygen will be available for the complete combustion of the gasoline. This is very important, as the burner, unless it is supplied with sufficient oxygen, will produce a low temperature, smoky flame. If the mechanic sees that the flame produced is yellow and deposits soot he will know that the burner is not obtaining sufficient oxygen. The flame should be a pale blue with a pink interior cone.

The vaporizing coil can now be wound. This is made up of copper or steel tubing with an internal diameter of 1/4 inch. Shelby steel tubing is more suitable for the vaporizing coil than copper, owing to the fact that it does not oxidize as rapidly and will, therefore, last much longer than copper. However, the Shelby steel tubing is not as easily bent as the copper. It will be necessary to wind the vaporizing coil over a wooden form to a diameter slightly larger than the outside diameter of the metal cylinder. The winding must be done on a lathe, and before the tube used is finally wound it should be filled with molten rosin to prevent it from flattening. If a larger helix was to be formed this would not be necessary, but with the one used on the burner the tubing will be sure to flatten out if this precaution is not taken. A piece several inches long is left extending from the coil so that it can be bent as shown in Fig. 40. The rosin can be removed from the helix by heating it thoroughly in a flame or fire and at the same time blowing through one end with a bicycle pump.

A small piece of brass about 1/2 inch wide, 2 inches long and 1/4 inch in thickness, is cut. The ends of this piece are slightly rounded so that the piece will fit in back of the burner. A hole is drilled through the center of this piece to accommodate the end of the vaporizing coil. The end of the vaporizing coil is then placed through this hole and the nipple is then soldered to the coil. This should be hard soldered as the operating temperature of the coil will very easily melt ordinary soft solder. When the soldering has been done the vaporizing coil is slipped over the metal cylinder previously made and the brass crosspiece riveted in place. This finishes the burner and it is now ready to be connected to the tank. A piece of flexible steel tubing is used to make the connection. Ordinary rubber tubing would not be suitable for this purpose as gasoline greatly shortens the life of rubber.

To operate the torch it is first necessary to heat the vaporizing coils to the proper temperature so that when the gasoline from the tank passes through them it will vaporize. This preliminary heating can be done by allowing a small trough of gasoline to burn under the coils, or the coils can be heated by the aid of another small blow torch of the hand variety. A pressure of about 10 pounds should be maintained in the tank. This pressure is indicated by the pressure gauge and although it is not necessary to know this pressure, at all times, the gauge is a very useful addition to the outfit. A heavy automobile pump is used to maintain the pressure in the tank.

A BRAZING FURNACE

When considerable brazing is to be done the worker should build a special furnace for that work alone, and, if possible, do the heating with gas. A blast of air will be necessary, but a very small blower, similar to that used for a portable forge, will do all that is required. The diagrams contained in Figs. 41 and 42 show plainly the construction of a small home-made furnace for brazing. This furnace may be built up on a bench with loose bricks, or it may be constructed in a more permanent manner, using an iron shell with fire brick lining instead of loose bricks.

For bench use, there is an insulating layer of bricks laid upon the wood, then an iron plate and then the

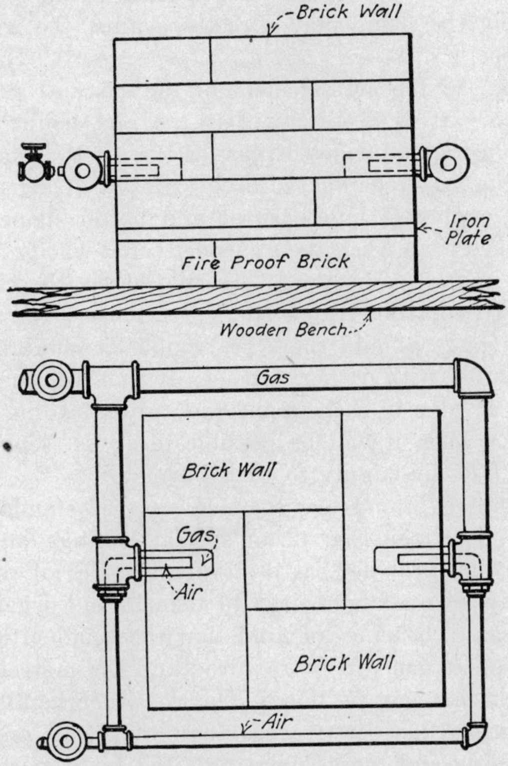

Figs. 41-42.—Details of the Brazing Furnace.

brick walls of the furnace are laid up. It is only necessary that the bricks are sufficient to hold the flame around the work to be brazed. There is a radiation or reflection of heat from a hot incandescent surface of brick or other material, like charcoal or carbon, which greatly aids in heating the object to be brazed. In fact, it is often impossible to braze certain work with the furnace at hand, until something has been placed around the work to keep the heat where it is needed—hence the use of the confined space in a brazing furnace instead of letting the flame play directly against the work, in the open.

In Fig. 42 the arrangement of the gas and air pipes is shown. It is necessary that the air should be delivered inside of the jet of gas, as the air thus supplied inside the gas, together with the supply of air outside the jet, enables a much better and hotter flame to be maintained than when the air is delivered wholly around the gas instead of inside it. In the sketch, the two jets are shown controlled by a single valve, each for the air and the gas. Should there be trouble in obtaining the best results with either burner, it can be cured by putting a valve in each of the four pipes leading to the burners. Then it will be possible to adjust separately, the gas and air supply to each burner.

The entire piping may be made up of standard fittings, as shown in Fig. 42, or special castings and forgings may be provided, as desired. The size of the furnace may be made sufficient to take in the usual work to be brazed, bearing in mind that the smaller the furnace, the less gas will be required and the more limited the work that can be done. On the other hand, while the large furnace costs more than the small one, and more piping and more gas is required to properly heat the apparatus there is always the possibility of filling the large furnace with bricks to fit it to small work and the gas can always be cut down to fit the furnace by means of the valves provided for that purpose. It is in order, then, to provide as large a furnace as there is likelihood of there being work for, then fill up the firepot with fire bricks until economy of gas is secured for the particular work to be done. Then, when a large piece of work comes along, take out the bricks, and a large furnace is at hand. The above remarks, of course, apply to job and repair work. For special manufacturing where the same work is to be done day after day, there will, of course, be provided special brazing furnaces, fitted for the particular work in hand.

TRANSFORMER FOR WELDING

A small heating transformer which can be used for welding the metals with lower melting points will be outlined in the following paragraphs.

In addition to the transformer described providing a high current density at the arc without causing an ex-

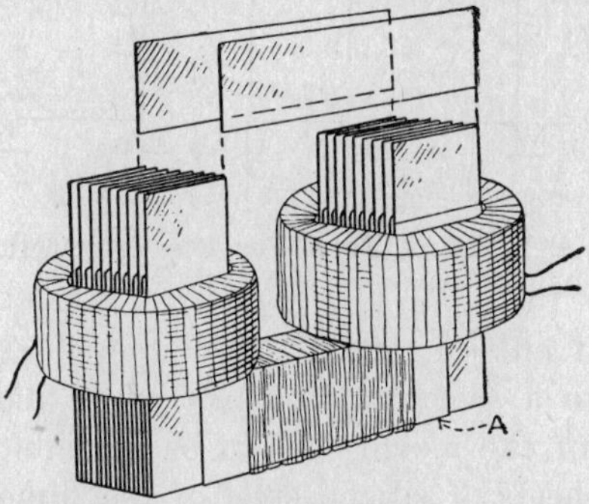

Fig. 43.—How the Final Pieces of the Transformer Core are Put in Place.

cessive overload on the line, good automatic regulation is secured with a machine of this type, and troubles due to short circuit when the arc is struck are completely eliminated. Since the coupling between the primary and secondary windings is made very loose, when the carbons are touched and the secondary short-circuited, heavy magnetic leakage takes place between the windings and the secondary voltage drops to practically zero. Upon separation of the electrodes, normal conditions are resumed and the leakage becomes very low. Regulation of the arc, should this be deemed necessary, is readily obtained by the interposition of a small bundle of transformer iron between the windings.

The transformer illustrated herewith was designed to operate on the regular 110 or 120-volt 60-cycle line. In the event that the machine is to be operated upon 25-cycle current the core must be made considerably larger than that shown, but the character of the windings need not be altered. The transformer takes a

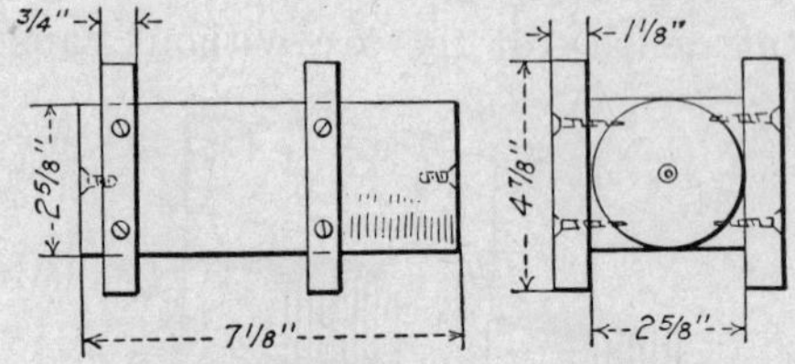

Fig. 44.—A Winding Form for the Transformer Coils.

primary current of about 15 amperes at full load, equivalent to a secondary output of 45 amperes.

The core of the machine is to be assembled first, and consists of sheets of silicon steel of the dimensions indicated in the drawings. About 25 pounds of this material are required for the complete core. The long sheets are laid one on another alternately projecting in either direction for a space equal to their width, and the ends of the strips comprising the short legs of the core are fitted in the ends of the long leg thus formed in a similar manner. This assembly is clearly depicted in the sketch, and is one which is soon mastered. Three legs of the core, two short and one long, are thus built up together, and all of them bound securely with two layers of friction tape. It is advisable to wear heavy gloves when working on the core, as the edges of the iron are exceedingly sharp. The core, when thus assembled, is ready for the windings.

Both the primary and secondary coils are wound upon the wooden form illustrated diagrammatically in the sketch. This may either be mounted between centers in the lathe and driven from the faceplate, or one end grasped in the chuck and the form steadied with the tailstock center for winding. A substantial winding jig

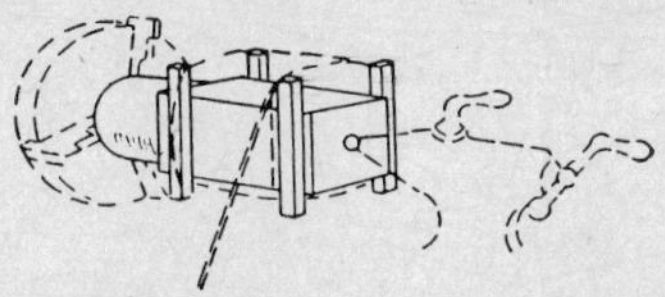

Fig. 45.—How the Winding Form is Mounted on the Lathe.

may be employed in the absence of a lathe, although the work is rather heavy.

The primary is wound with 150 turns of No. 14 cotton covered wire. The winding form was first wrapped with a layer of heavy cord between the cross bars and a wrapping of insulating paper secured over the cord. One end of the wire is secured to the form and a layer of wire wound on the form, running the lathe backwards and at its lowest speed. The wire should be free from kinks and fed from a spool under constant tension. Considerable care must be exercised to get the turns close and even. Any imperfections in the insulation of the wire may be covered with empire cloth wrapped around the wire at that point.

After the first layer is completed, it is covered with insulating paper and a second layer wound over it, repeating the procedure until as many layers as are necessary to secure the 150 turns are in place. The form is then removed from the lathe, and, maintaining a hold on the coil, the two crosspieces on one end of the form are removed. The layer of cord under the wire may now be pulled out and a few pieces of cord passed through the coil and tied around it, holding the wire in place temporarily. The coil is now lifted off the form and wound with friction tape, the ends of the winding are soldered to flexible leads, the soldered joints

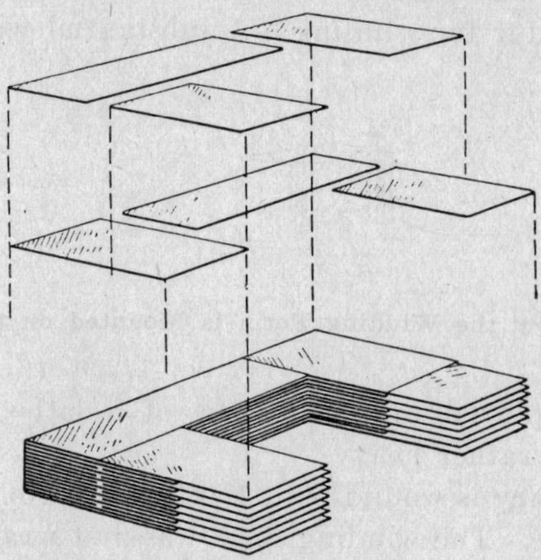

Fig. 46.—How the Transformer Core is Assembled.

being insulated first with empire cloth and then taped to the coil. The winding is now ready to be placed on the core.

The secondary coil was wound in precisely the same way, with the exception that two No. 8 wires in parallel were employed—40 turns of this double wire, or 80 turns in all, are required. Since wire of this size is far more flexible, the turns may have to be tapped lightly with a mallet to get them in place at the ends of the layers. The ends of the winding are soldered to heavy leads which are fitted with lugs at the terminals for connection to the electrodes of the furnace.

Stranded No. 4 wire rubber covered is suitable for these leads, though a somewhat more flexible connection is preferable. After thorough taping, both primary and secondary windings and the core are given a coat of shellac.

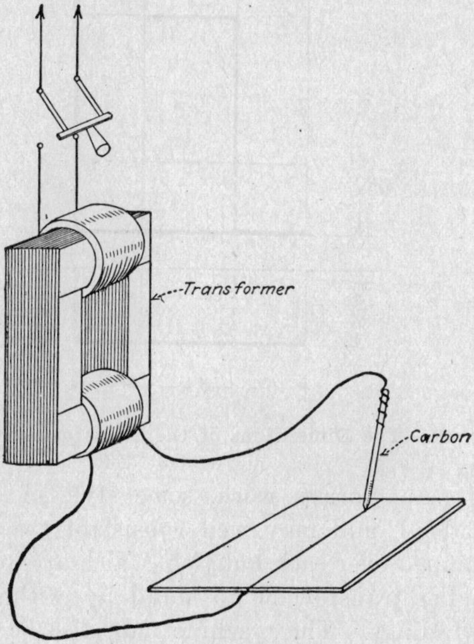

Fig. 47.—The Connections for the Transformer.

The windings should be a loose fit on the short legs of the core, and are held in place by small wooden wedges driven in on opposite sides. With the windings in place, the magnetic circuit may be completed by insertion of the strips comprising the fourth leg of the core. The ends of the projecting short legs are spread one by one and the long strips inserted, being "staggered" exactly as were the other strips. In this case, however, the two outside strips are placed so as to project half-way over each end strip, being centered on the core, and a wrapping of tape and coat of shellac applied. While it is not essential to mount the trans-

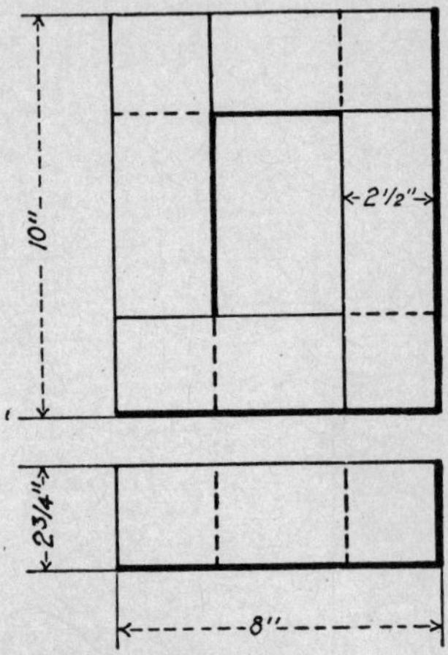

Fig. 48.—The Dimensions of the Transformer Core.

former for temporary work, some style of support is usually added, and may well consist of two hardwood strips clamped over each long leg, which are long enough to allow the transformer to stand up without resting on the windings. The machine may also be laid down flat if desired.

PART V

SOLDERING NOTES

Granite and enameled ware can be soldered by first scraping off the rough edges of enamel around the hole. This should be done for a distance of 1/4 inch or more depending somewhat upon the size of the leak. The metal on the under side of the leak should be brightened with emery cloth. The pan is then turned right side up and hot soldering copper held to it underneath the hole until a stick of solder will melt when placed on the

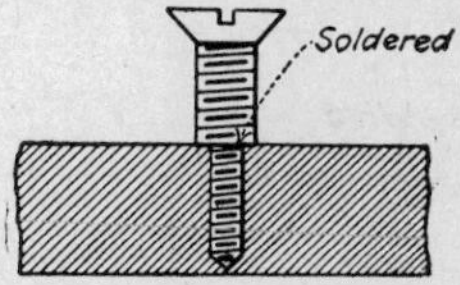

Fig. 48A.—How a Broken Machine Screw Can be Removed.

inside. In this way the solder is caused to melt and run into the hole. When this happens the copper should be taken away to permit the solder to cool. In doing work of this kind the mechanic should learn that all soldering must be done on the inside of the pan with sufficient moisture to keep it from melting. When there is water in a pan the heat merely travels through the solder to the water and therefore the solder is prevented from reaching the melting point.

A broken machine screw can be removed by soldering another piece to it, as illustrated in Fig. 48A. The top surface of the broken screw is carefully cleaned and a drop of solder placed upon it. The solder is then heated by bringing the copper in proximity to it and a machine screw is then soldered to the broken screw by bringing it in contact and allowing it to cool.

A good substitute for a wiping joint is shown in Fig. 49. The ordinary plumber's method of wiping a joint requires much experience and the novice is advised not to attempt such a job without knowing something about it. The method shown in Fig. 49 will produce a good

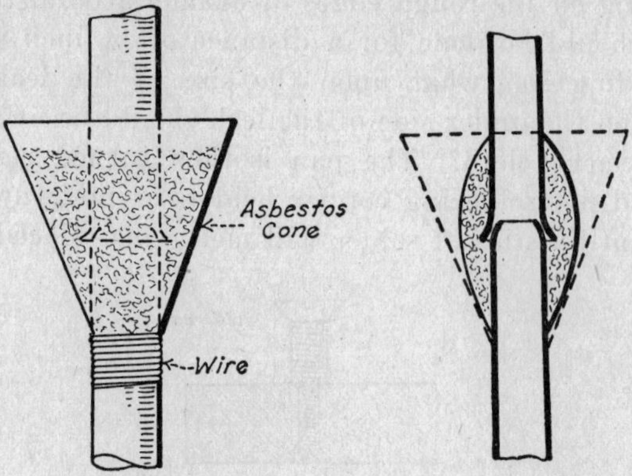

Fig. 49.—A Substitute for a Wiped Joint.

joint for most purposes. A cone-shaped piece of asbestos is fastened to the pipe with wire and molten solder is poured into the top. When this is cool the asbestos is removed and the surplus solder is taken off with a wood rasp.

A soldering copper that will maintain a continuous heat is shown in Fig. 50. This is very good for use in a shop where continuous soldering is necessary, such as in making electrical connections. The heat at that point of the copper can be regulated by moving the burner toward the point for more heat or toward the back of the copper for less.

Oftentimes the threads of a water faucet become

stripped and cause the top of the faucet to leak. To remedy this the threaded part of the faucet is removed, cleaned and well tinned with a hot soldering copper. After the threads have been thoroughly cleaned with a thin layer of solder the threaded portion is screwed

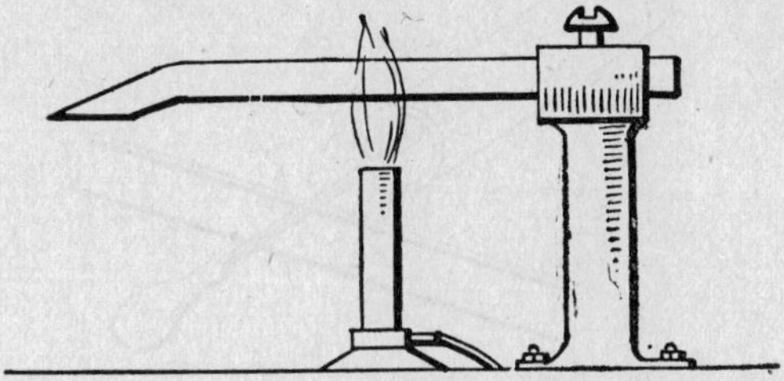

Fig. 50.—A Soldering Copper That Will Maintain a Continuous Heat.

back into the faucet. It will be found that until the faucet is worked a little it may be difficult to screw the piece back into place. However, this method will remedy the leak and put the faucet in usable condition.

A broken tooth in a ratchet wheel can be very easily repaired by the method shown in Fig. 51. In this case

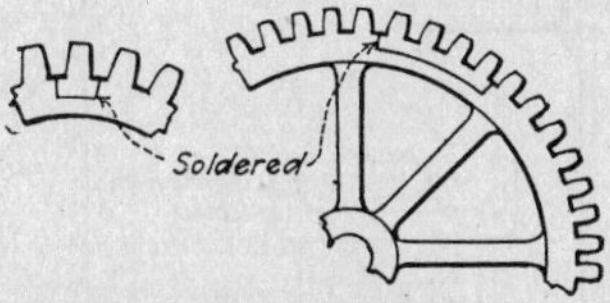

Fig. 51.—Replacing a Broken Tooth in a Ratchet Wheel.

the old tooth is filed completely out and the space for the new one is made. A new tooth is then cut from metal and soldered into place. If several teeth are stripped off a piece can be filed with several teeth in it and soldered into place in the same way as the one tooth was soldered into place.

A very good method of heating a soldering copper is

shown in Fig. 52. This is in case the copper is heated in a hot coal fire. If the copper is allowed to come in contact with the hot coal the tinning will oxidize and the copper will be unsuitable for use when withdrawn

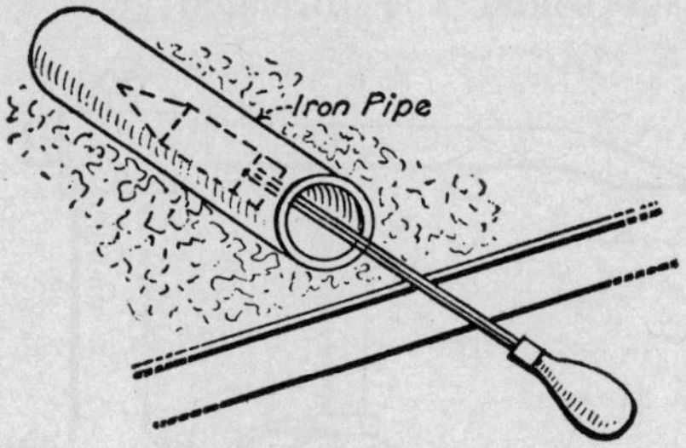

Fig. 52.—How a Soldering Copper Should be Heated in a Coal Fire.

from the fire. By slipping the copper into a piece of iron pipe the heating can be accomplished without danger of destroying the tinning.

A good method of tinning copper or brass wire is shown in Fig. 53. Considerable heat must be used under

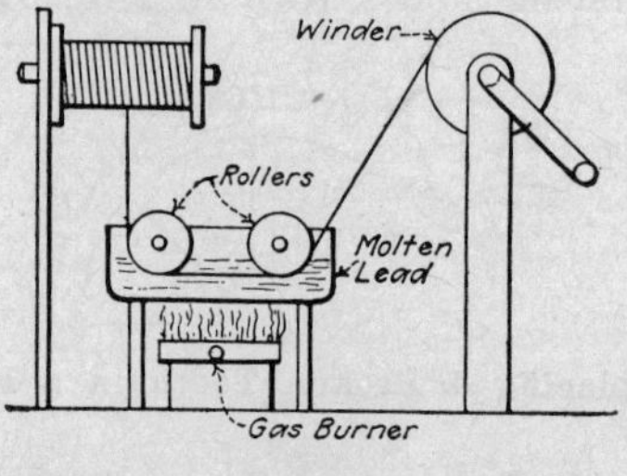

Fig. 53.—Simple Apparatus for Tinning Brass or Copper Wires.

the pot of solder to keep the solder in a thoroughly molten condition. In fact it may be necessary to experiment somewhat with the heat of the solder as there is just the right temperature where no trouble will be had in making it adhere to the wire. If it is too cool it will not adhere and if it is very hot probably the same trouble will be had. The apparatus used is so simple that the mechanic will gather the details from the drawing.

The American Steel and Wire Company explains a method of making a very strong stay wire fastening: "The short end of the strand, after passing through the eye, is temporarily tied to the main part with string or wire if necessary (see Fig. 54). Tough annealed iron wire or soft brass wire used for sizing is first laid into the groove between the two parts of strand. About three inches from the eye, the sizing wire is given a right angle

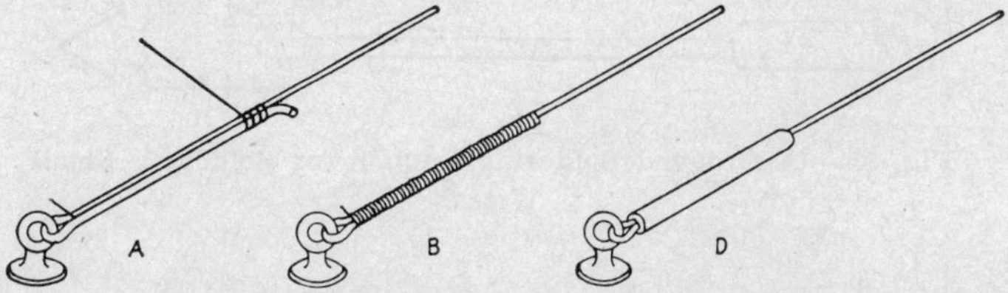

Fig. 54.—Method of Making a Strong Stay Wire Fastening.

bend and the wrapping begun (a); the ends of the sizing wire are twisted together (b) and laid against the sizing (c). The wires in the short projecting end of strand are next loosened or opened by pinching with pliers (c). This is done in order that the solder may adhere to the wires and form a knob that cannot pull out of the sizing. The entire sized fastening is then cleaned with benzine, coated with soldering paste, and heavily soldered (d). If the surface of tinned or galvanized stay wire or strand has been scratched in securing it to eye-bolts, rust spots will soon appear, especially as the moisture settling on the stay runs down and collects on the fastening itself. It is therefore a wise precaution to paint all stay fastenings with black asphaltum paint or turpentine japan."

A writer in Telephony tells of his very good grooved copper for soldering heavy wires: "I had some trouble in soldering small insulated wires, usually of No. 18 B & S Gauge to line wires. To make the joint I was in the habit of removing the insulation from the small wire over a length of six or eight inches and winding it tightly around the larger wire after cleansing both wires thoroughly with sandpaper. Usually a good hot copper, applied to this with proper flux would make the solder stick, but if solder was applied before the joint was hot, much ran to the lower side of the copper and dripped off, or if the work was in some inconvenient

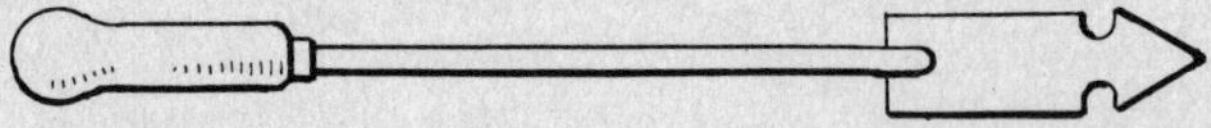

Fig. 55.—A Grooved Soldering Copper for Soldering Small Wires.

place the copper often was too much cooled before we could get in position. Under these conditions solder often sticks only on one side, or in scattered spots.

I cured my trouble by filing a small groove across my copper just back of the bevel. This groove is about $3/16$ inch deep and of like width. A good groove can be easily cut in a copper by clamping a flat piece of iron (buggy tire is good) against the face of the copper in a vise, and drilling between the copper and the plate of iron. I find a $3/16$ inch or $9/23$ inch drill about the right size for a No. 12 wire. The drill should be made to cut to at least two-thirds of its diameter into the copper.

A special copper for use in soldering commutators is shown in Fig. 56. In place of the ordinary soldering copper with the usual 60 degree angle a blunt pointed copper with about 80 degrees keeps the heat at the point where it is most needed. Those familiar with commutator work will appreciate the usefulness of this idea.

A convenient little torch for soldering where there is a wind or draught is illustrated in Fig. 57. It consists

Fig. 56.—A Special Soldering Copper for Work on Commutators.

merely of a candle set in a small cylindrical tin can. Two slots are cut to accommodate the wire.

In soldering polished lugs and well kept switchboards the electrician generally is troubled with the solder running and adhering to the outside of the lug. To prevent this nuisance a small piece of asbestos sheet is tied around the top of the lug with wire and in this way

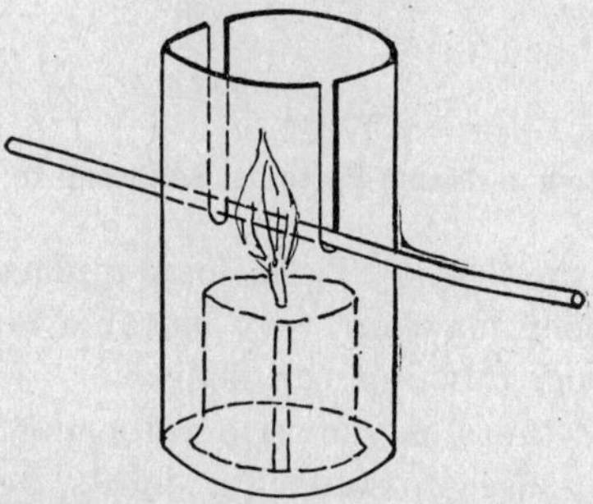

Fig. 57.—A Little Torch for Soldering Small Wires.

the soldering can be carried on in the usual way without the solder adhering to the polished part of the lug.

A very easy method of soldering a name plate on to a circular piece of work is illustrated in Fig. 58. In this case the cylinder was polished and nickel-plated and therefore the soldering must be done without causing discoloration. The plate was first rolled to conform with the circumference of the cylinder. The back of the plate was spread very carefully with a soldering flux. Over this was sprinkled the soldering filings prepared especially for this purpose. The name plate was then placed in position and held by a brass band arranged with a clasp. The heat was then applied to the band and allowed to penetrate the band and melt the solder. In this way a nice clean job was made possible.

When electricians are working in a house with finished floors they often have trouble with the solder spattering

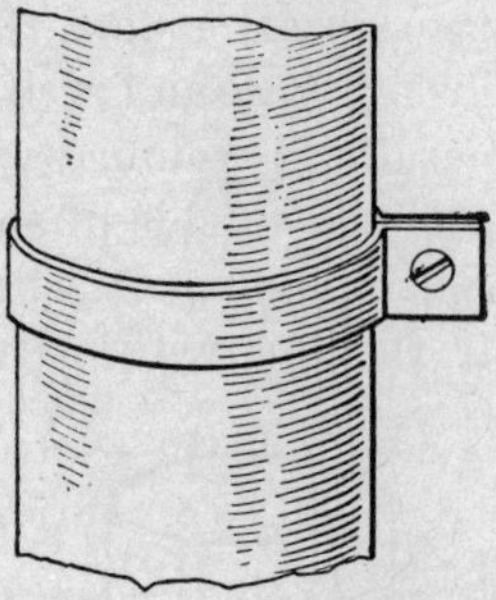

Fig. 58.—How a Name-Plate is Soldered to a Cylinder.

and burning the floors. Tissue paper placed over heavy wrapping paper makes a very suitable mat to prevent the solder from burning the floor.

In Fig. 59 there is shown a soldering copper made especially for use in soldering joints near a ceiling. The ordinary soldering copper can be easily converted into one of the type shown. It will be seen that the use of this copper will eliminate the possibility of burning one's hands while doing work overhead.

Badly corroded copper wires can be brought to condition for soldering by removing the insulation and immersing the ends of the wire in a solution of nitric acid 25% and water 75%. The end of the wires are then washed in water and allowed to dry. This method will make the ends of the wires bright and clean and ready to receive the solder.

A writer in Telephony used the following method for rusty telephone joints: "In less than six months on one small telephone exchange practically every joint

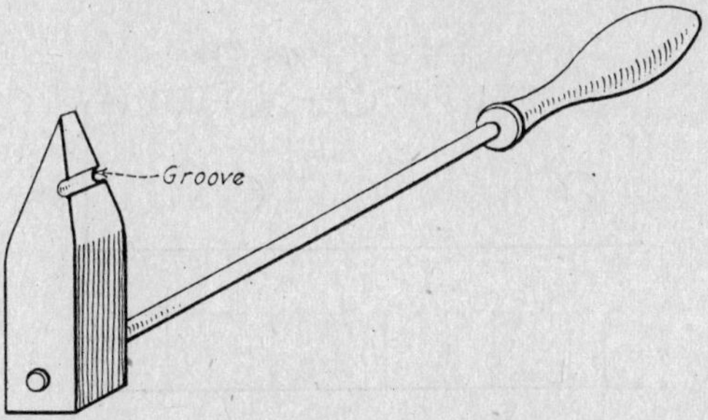

Fig. 59.—A Soldering Copper Used Especially in Making Joints Near a Ceiling.

in a 14-mile metallic line had rusted out. New joints were promptly cut in, but in a few months these were gone. The trouble was found to be due to the use of soldering acid when making the joint, and the acid was not properly washed off. Galvanized iron will last just

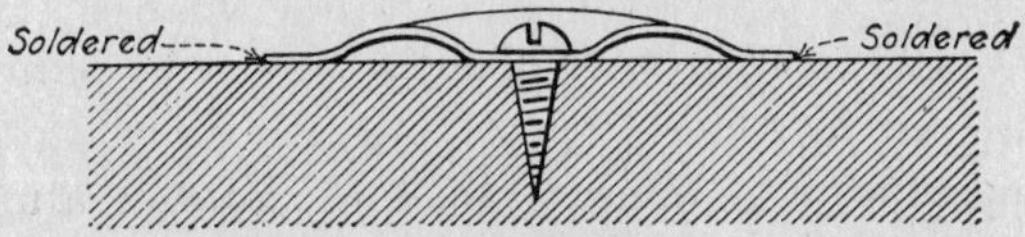

Fig. 60.—Method of Stopping Leaks on a Tin Roof.

so long as the galvanizing lasts. As soon as the zinc coating or galvanizing is penetrated, the iron begins to rust and is soon eaten away. Every man who has put up a wire fence knows this. Yet these same men, when putting up a telephone line, will bruise the galvanizing at each tie, or will use acid when soldering the joints."

A convenient and reliable method of stopping a leak in a tin roof is shown in Fig. 60. In place of employing a nail, a screw is used owing to the fact that it is not possible to pull it up as easily as a nail. After the disc is screwed down the solder is flowed over the head of the screw to the disc and over the edge of the disc to the

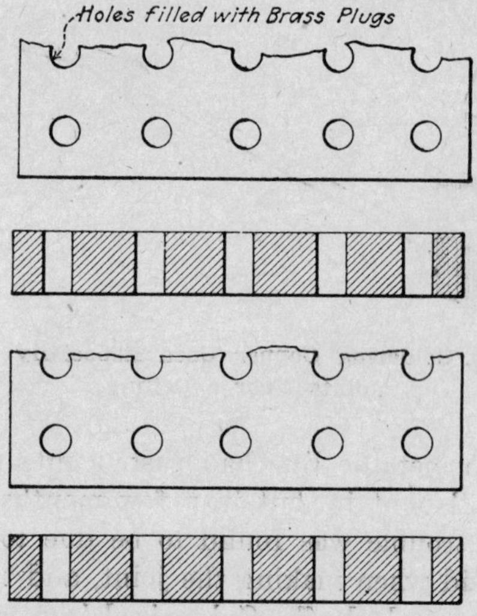

Fig. 61.—Method of Soldering Aluminum.

tin roof. It is well to give the disc a coat of paint after it is soldered in place.

A somewhat ingenious method of joining aluminum pieces together with ordinary solder is shown in Fig. 61. The holes are first drilled in the aluminum and these holes are then filled with brass plugs. After the plugs are driven in place they should be filed off flush with the surface of the aluminum. By flowing solder over the plugs and both pieces of the aluminum it will be found possible to sweat the pieces together in the usual way. It will be understood that aluminum cannot be soldered successfully by the ordinary process, owing to the fact that a thin film of oxide forms so quickly that it is impossible to apply the solder to the clean metallic surface. The surface of aluminum may look entirely clean to the naked eye, but there is always present this film of aluminum oxide.

A writer in Popular Mechanics states that he often used the hot engine of his motorcycle as a soldering iron

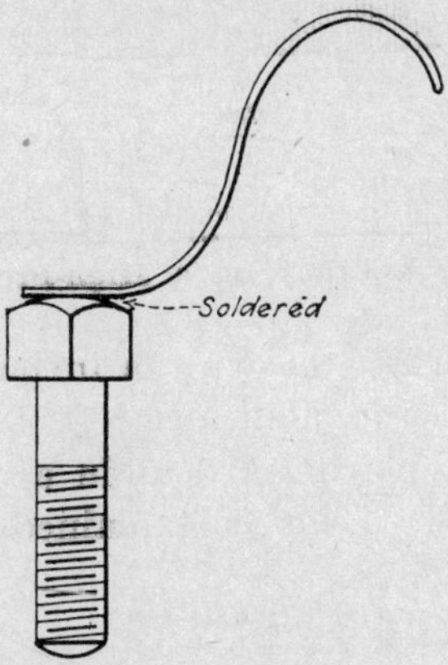

Fig. 62.—Method of Placing a Bolt in an Inaccessible Locality.

heater in his work as a "telephone trouble shooter." He laid the copper wires to be soldered upon the cylinder and allowed them to remain there until they became hot enough to melt the solder. The same writer also states that he lighted his alcohol torch with the spark from the spark coil by touching it to the frame of the motorcycle.

A very efficient method of placing a bolt in a difficult locality is shown in Fig. 62. A small strip of metal is soldered to the bolt and this can be bent in any convenient position so that the bolt can be replaced in an inaccessible hole.

Fig. 63 shows an ingenious method of repairing a crack in a high pressure steam pipe. The pipe is first thoroughly cleaned for a distance of one inch beyond each side of the crack. This is then wound over with

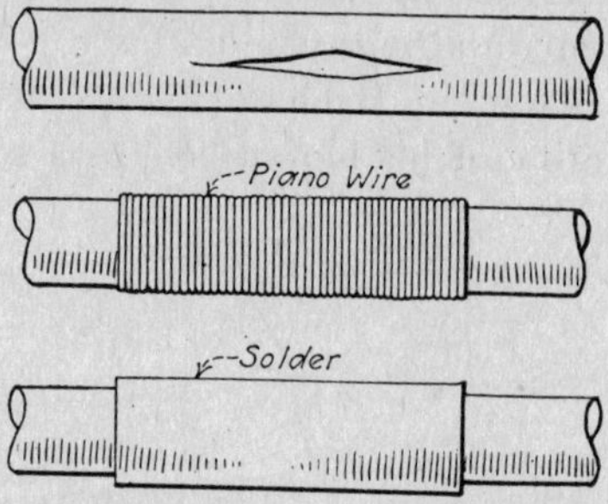

Fig. 63.—Repairing a Crack on a High-Pressure Steam Pipe.

piano wire and when the wire is in position the solder is flowed around the joint.

A method of repairing a dent in a metal cylinder is shown in Fig. 64. This is explained by a writer in

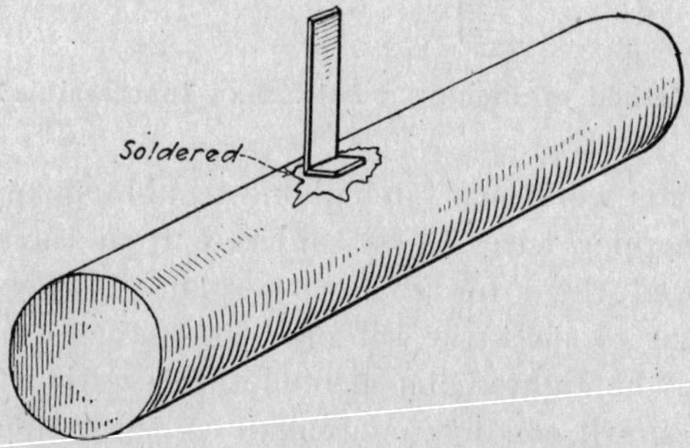

Fig. 64.—Taking a Dent Out of a Metal Cylinder.

Popular Mechanics: Dents in any kind of sheet metal, especially the brass fixtures of an automobile, are exceedingly difficult to straighten out and make a neat appearing piece of work. One of the best methods of doing this is to make a clip of metal, steel, or brass as shown at A in the sketch, and solder the small part turned at right angles to the metal in the dent. The sunken metal can then be pulled up, the clip unsoldered and the surface polished. It is not necessary to drill a hole in the fixture, and the surface can be smoothed up like new.

In soldering electrical connections to posts set in hard rubber such as in automobiles, the rubber can be prevented from melting by holding the plug with a wet cloth. This is to prevent too much heat from reaching the rubber.

INDEX

1920

REVISED

CATALOGUE

of LATEST *and* BEST

Practical and Mechanical Books

Including Automobile and Aviation Books

PRACTICAL BOOKS FOR PRACTICAL MEN

Any of these books will be sent prepaid to any part of the world, on receipt of price. Remit by Draft, Postal Order, Express Order or Registered Letter.

Published and For Sale by

The Norman W. Henley Publishing Co.

2 West 45th Street New York, U. S. A.

INDEX TO SUBJECTS

☞ **Any of these books will be sent prepaid to any part of the world, on receipt of price.**

REMIT by Draft, Postal Money Order, Express Money Order, or by Registered Mail.

GOOD, USEFUL BOOKS

AUTOMOBILES—MOTORCYCLES

The Modern Gasoline Automobile, Its Design, Construction, Operation.

By VICTOR W. PAGÉ, M.S.A.E. This is the most complete, practical, and up-to-date treatise on gasoline automobiles and their component parts ever published. In the new revised and enlarged 1920 edition, all phases of automobile construction, operation and maintenance are fully and completely described and in language anyone can understand. Every part of all types of automobiles, from light cyclecars to heavy motor trucks and tractors, are described in a thorough manner; not only the automobile, but every item of its equipment, accessories, tools needed, supplies and spare parts necessary for its upkeep, are fully discussed. It is clearly and concisely written by an expert familiar with every branch of the automobile industry and the originator of the practical system of self-education on technical subjects; it is a liberal education in the automobile art, useful to all who motor for either business or pleasure. Anyone reading the incomparable treatise is in touch with all improvements that have been made in motor car construction. All latest developments, such as high speed aluminum motors and multiple valve and sleeve valve engines, are considered in detail. The latest ignition, carburetor and lubrication practice is outlined. New forms of change speed gears, and final power transmission systems, and all latest chassis improvements, are shown and described. This book is used as a text in all leading automobile schools, and is conceded to be the standard treatise. The chapter on Starting and Lighting Systems has been greatly enlarged, and many automobile engineering features that have long puzzled laymen are explained so clearly that the underlying principles can be understood by anyone. This book was first published six years ago, and so much new matter has been added to the book that it is nearly twice its original size. The only treatise covering various forms of war automobiles and recent developments in motor truck design, as well as pleasure cars. This book is not too technical for the layman nor too elementary for the more expert. It is an incomparable work of reference for home or school. 6 x 9. Cloth, 1,000 pages, nearly 1,000 illustrations, 12 folding plates.

Price, $4.00

Questions and Answers Relating to Modern Automobile Construction, Driving and Repair.

By VICTOR W. PAGÉ. A self-educator on automobiling without an equal. This practical treatise consists of a series of thirty-seven lessons, covering with over 2,000 questions and their answers—the automobile, its construction, operation and repair. The subject matter is absolutely correct and explained in simple language. If you can't answer all of the following questions, you need this work. The answers to these and 2,000 more are to be found in its pages.

Give the name of all important parts of an automobile and describe their functions. Describe action of latest types of kerosene carburetors. What is the difference between a "double" ignition system and a "dual" ignition system? Name parts of an induction coil. How are valves timed? What is an electric motor starter and how does it work? What are advantages of worm drive gearing? Name all important types of ball and roller bearings. What is a "three-quarter" floating axle? What is a two-speed axle? What is the Vulcan electric gear shift? Name the causes of lost power in automobiles. Describe all noises due to deranged mechanism and give causes. How can you adjust a carburetor by the color of the exhaust gases? What causes "popping" in the carburetor? What tools and supplies are needed to equip a car? How do you drive various makes of cars? What is a differential lock and where is it used? Name different systems of wire wheel construction. What is a "positive" drive differential? etc., etc. Answers every question asked relating to the modern automobile. A popular work at a popular price. 5¼ x 7½. Cloth, 650 pages, 392 illustrations, 3 folding plates. 1920 Edition just published.

Price, $2.50

How to Run an Automobile.

By VICTOR W. PAGÉ. This treatise gives concise instructions for starting and running all makes of gasoline automobiles, how to care for them, and gives distinctive features of control. Describes every step for shifting gears, controlling engine, etc. Among the chapters contained are: I. Automobile Parts and Their Functions. II. General Starting and Driving Instructions. III. Typical 1919 Control Systems—Care of Automobiles. Thoroughly illustrated. 178 pages, 72 illustrations.

Price, $1.50

The Automobilist's Pocket Companion and Expense Record.

By VICTOR W. PAGÉ. This book is not only valuable as a convenient cost record, but contains much information of value to motorists. Includes a condensed digest of auto laws of all States, a lubrication schedule, hints for care of storage battery and care of tires, location of road troubles, anti-freezing solutions, horsepower table, driving hints and many useful tables and recipes of interest to all motorists. Not a technical book in any sense of the word, just a collection of practical facts in simple language for the everyday motorist. Convenient pocket size.

Price, $1.25

Gasoline and Kerosene Carburetors, Construction, Installation and Adjustment.

By Capt. V. W. Pagé. All leading types of carburetors are described in detail, special attention being given to the forms devised to use the cheaper fuels such as kerosene. Carburetion troubles, fuel system troubles, carburetor repairs and installation, electric primers and economizers, hot spot manifolds and all modern carburetor developments are considered in a thorough manner. Methods of adjusting all types of carburetors are fully discussed as well as suggestions for securing maximum fuel economy and obtaining highest engine power. 250 pages, 89 illustrations. **Price, $2.00**

Starting, Lighting and Ignition Systems.

By Victor W. Pagé. A practical treatise on latest automobile starting, lighting and ignition system practice. This practical volume has been written with special reference to the requirements of the reader desiring easily understood explanatory matter relating to all types of automobile electrical systems. It can be understood by anyone, even without electrical knowledge, because elementary electrical principles are considered before any attempt is made to discuss features of the various systems. These basic principles are clearly stated and illustrated with simple diagrams. All the leading systems of starting, lighting and ignition have been described and illustrated with the cooperation of the experts employed by the manufacturers. Over 200 wiring diagrams are shown in both technical and nontechnical forms. Complete data is given for locating troubles in all systems, the various steps being considered in a logical way for those without expert electrical knowledge. All ignition systems receive full consideration, starting with the simplest battery and coil forms found on early cars to the modern short-contact timer and magneto methods used with the latest eight and twelve-cylinder motors. Full directions are given for making all repairs and complete instructions for locating troubles with meters. This book is unusually complete, as it also includes descriptions of various accessories operated by electric current, such as electrical gear shifts, brake actuation, signaling devices, vulcanizers, etc. Over 700 pages. 425 specially made engravings. 1920 Edition, Revised and Enlarged. **Price, $3.00**

Automobile Welding with the Oxy-Acetylene Flame.

By M. Keith Dunham. Explains in a simple manner apparatus to be used, its care, and how to construct necessary shop equipment. Proceeds then to the actual welding of all automobile parts, in a manner understandable by everyone. *Gives principles never to be forgotten.* This book is of utmost value, since the perplexing problems arising when metal is heated to a melting point are fully explained and the proper methods to overcome them shown. 167 pages, fully illustrated. **Price, $1.50**

Automobile Repairing Made Easy.

By Victor W. Pagé. A thoroughly practical book containing complete directions for making repairs to all parts of the motor car mechanism. Written in a thorough but non-technical manner. Gives plans for workshop construction, suggestions for equipment, power needed, machinery and tools necessary to carry on business successfully. Tells how to overhaul and repair all parts of all automobiles. The information given is founded on practical experience, everything is explained so simply that motorists and students can acquire a full working knowledge of automobile repairing. Other works dealing with repairing cover only certain parts of the car—this work starts with the engine, then considers carburetion, ignition, cooling and lubrication systems. The clutch, change speed gearing and transmission system are considered in detail. Contains instructions for repairing all types of axles, steering gears and other chassis parts. Many tables, short cuts in figuring and rules of practice are given for the mechanic. Explains fully valve and magneto timing, "tuning" engines, systematic location of trouble, repair of ball and roller bearing, shop kinks, first aid to injured and a multitude of subjects of interest to all in the garage and repair business. All illustrations are especially made for this book, and are actual photographs or reproductions of engineering drawings. This book also contains Special Instructions on Electric Starting, Lighting and Ignition Systems, Tire Repairing and Rebuilding, Autogenous Welding, Brazing and Soldering, Heat Treatment of Steel, Latest Timing Practice, Eight and Twelve-Cylinder Motors, etc., etc. You will never "Get Stuck" on a Job if you own this book. 1,000 specially made engravings on 500 plates. 1,056 pages (5½ x 8). 1920 Edition. **Price, $4.00**

The Model T Ford Car, Its Construction, Operation and Repair, Including the Fordson Farm Tractor, F. A. Lighting and Starting System, Ford Motor Truck.

By Victor W. Pagé. This is the most complete and practical instruction book ever published on the Ford car and Fordson tractor. A high grade, cloth bound book, printed on the best paper, illustrated by specially made drawings and photographs. All parts of the Ford Model T car and Fordson tractor are described and illustrated in a comprehensive manner. The construction is fully treated and operating principle made clear to everyone. Complete instructions for driving and repairing are given. To the 1920 Revised Edition matter has been added on the Ford Truck and Tractor Conversion Sets and Genuine Fordson Tractor. All parts are described. All repair processes illustrated and fully explained. Written so all can understand—no theory, no guesswork. New Edition. 153 illustrations, 410 pages, 2 large folding plates. **Price, $2.00**

Motorcycles, Side Cars and Cyclecars, Their Construction, Management and Repair.

By Victor W. Pagé. Describes fully all leading types of machines, their design, construction, maintenance, operation and repair. 550 pages. 350 specially made illustrations, 5 folding plates. New Edition. **Price, $2.50**

Automobile Charts

By VICTOR W. PAGÉ, M.S.A.E.

THE POPULAR AUTOMOBILE SERIES
UNIFORM SIZE—24"x 38"—PRICE 35 CENTS EACH

Location of Gasoline Engine Troubles Made Easy.

This chart shows clearly all parts of a typical four-cylinder gasoline engine of the four-cycle type. It simplifies location of all engine troubles. No details omitted. **Price, 35 cents**

Location of Carburetion Troubles Made Easy.

It shows clearly how to find carburetion troubles and names all defects liable to exist in the various parts. Instructions are given for carburetor adjustment. **Price, 35 cents**

Location of Ignition System Troubles Made Easy.

In this chart all parts of a typical double ignition system using battery and magneto current are shown, and suggestions are given for readily finding ignition troubles and eliminating them when found. **Price, 35 cents**

Location of Cooling and Lubricating Troubles.

This is a combination chart showing all components of the approved form of water cooling group as well as a modern engine lubrication system. It shows all points where defects exist that may result in engine overheating, both in cooling and oiling systems. **Price, 35 cents**

Lubrication of the Motor Car Chassis.

This chart presents the plan view of a typical six-cylinder chassis of standard design and outlines all important bearing points requiring lubrication, and is a valuable guide to the correct lubrication of any modern car. A practical chart for all interested in motor car maintenance. **Price, 35 cents**

While each chart is complete in itself, the set covers all maintenance instructions for the entire automobile. Sold singly. Securely wrapped.

Location of Starting and Lighting System Faults.

The most complete chart yet devised, showing all parts of the modern automobile starting, lig ting and ignition systems, giving instructions for systematic location of all faults in wiring, lamps, motor or generator, switches and all other units. Invaluable to motorists, chauffeurs and repairmen. Size 24 x 38 inches. **Price, 35 cents**

Location of Ford Engine Troubles Made Easy.

Chart showing clear sectional views depicting all portions of the Ford power plant and auxiliary groups. It outlines clearly all parts of the engine, fuel supply systems, ignition group and cooling system, that are apt to give trouble, detailing all derangements that are liable to make an engine lose power, start hard, or work irregularly. This chart simplifies location of all engine faults. Size 25 x 38 inches. **Price, 35 cents**

Location of Motorcycle Troubles Made Easy.

This chart simplifies location of all power-plant troubles and will prove of value to all who have to do with the operation, repair or sale of motorcycles. No details omitted. Size 30 x 20 inches. **Price, 35 cents**

AVIATION

A B C of Aviation.

By CAPT. V. W. PAGÉ. This book describes the basic principles of aviation, tells how a balloon or dirigible is made and why it floats in the air. Describes how an airplane flies. It shows in detail the different parts of an airplane, what they are and what they do. Describes all types of airplanes and how they differ in construction; as well as detailing the advantages and disadvantages of different types of aircraft. It includes a complete dictionary of aviation terms and clear drawings of leading airplanes. The reader will find simple instructions for unpacking, setting up and rigging airplanes. A full description of airplane control principles is given and methods of flying are discussed at length.

This Book answers every question one can ask about modern aircraft, their construction and operation. A self educator on aviation without an equal. 275 pages, 130 specially made illustrations with 7 plates. **Price, $2.50**

Aviation Engines—Design; Construction; Repair.

By LIEUT. VICTOR W. PAGÉ, Aviation Section, S.C.U.S.R. This treatise, written by a recognized authority on all of the practical aspects of internal combustion engine construction, maintenance and repair, fills the need as no other book does. The matter is logically arranged; all descriptive matter is simply expressed and copiously illustrated, so that anyone can understand airplane engine operation and repair even if without previous mechanical training. This work is invaluable for anyone desiring to become an aviator or aviation mechanician.

The latest rotary types, such as the Gnome Monosoupape, and LeRhone, are fully explained, as well as the recently developed Vee and radial types. The subjects of carburetion, ignition, cooling and lubrication also are covered in a thorough manner. The chapters on repair and maintenance are distinctive and found in no other book on this subject. Not a technical book, but a practical, easily understood work of reference for all interested in aeronautical science. 576 pages, 253 illustrations. **Price, Net, $3.00**

Glossary of Aviation Terms—English-French; French-English.

A complete glossary of practically all terms used in aviation, having lists in both French and English with equivalents in either language compiled by LIEUTS. VICTOR W. PAGÉ, A.S., S.C.U.S.R., and PAUL MONTARIOL, of the French Flying Corps. **Price, Net, $1.00**

Aviation Chart—Location of Airplane Power Plant Troubles Made Easy.

By LIEUT. VICTOR W. PAGÉ, A.S., S.C.U.S.R. A large chart outlining all parts of a typical airplane power plant, showing the points where trouble is apt to occur and suggesting remedies for the common defects. Intended especially for aviators and aviation mechanics on school and field duty. **Price. 35 cents**

BRAZING AND SOLDERING

Brazing and Soldering.

By JAMES F. HOBART. The only book that shows you just how to handle any job of brazing or soldering that comes along; it tells you what mixture to use, how to make a furnace if you need one. Full of valuable kinks. The fifth edition of this book has just been published, and to it much new matter and a large number of tested formulas for all kinds of solders and fluxes have been added. **Price, 35c.**

CHARTS

Aviation Chart—Location of Airplane Power Plant Troubles Made Easy.

By LIEUT. VICTOR W. PAGÉ, A.S., S.C.U.S.R. A large chart outlining all parts of a typical airplane power plant, showing the points where trouble is apt to occur and suggesting remedies for the common defects. Intended especially for aviators and aviation mechanics on school and field duty. **Price, 35 cents**

Modern Submarine Chart—With 200 Parts Numbered and Named.

A cross-section view, showing clearly and distinctly all the interior of a submarine of the latest type. No details omitted—everything is accurate and to scale. This chart is really an encyclopedia of a submarine. **Price, 25 cents**

Box Car Chart.

A chart showing the anatomy of a box car, having every part of the car numbered and its proper name given in a reference list. **Price, 25 cents**

Gondola Car Chart.

A chart showing the anatomy of a gondola car, having every part of the car numbered and its proper reference name given in a reference list. **Price, 25 cents**

Passenger Car Chart.

A chart showing the anatomy of a passenger car, having every part of the car numbered and its proper name given in a reference list. **Price, 25 cents**

Steel Hopper Bottom Coal Car.

A chart showing the anatomy of a steel hopper bottom coal car, having every part of the car numbered and its proper name given in a reference list. **Price, 25 cents**

Tractive Power Chart.

A chart whereby you can find the tractive power or drawbar pull of any locomotive without making a figure. Shows what cylinders are equal, how driving wheels and steam pressure affect the power. What sized engine you need to exert a given drawbar pull or anything you desire in this line. **Price, 50 cents**

Horse-power Chart.

Shows the horse-power of any stationary engine without calculation. No matter what the cylinder diameter of stroke, the steam pressure or cut-off, the revolutions, or whether condensing or non-condensing, it's all there. Easy to use, accurate and saves time and calculations. Especially useful to engineers and designers. **Price, 50 cents**

Boiler Room Chart.

By George L. Fowler. A chart—size 14 x 28 inches—showing in isometric perspective the mechanisms belonging in a modern boiler room. This chart is really a dictionary of the boiler room—the names of more than 200 parts being given.

Price, 25 cents

COKE

Coke—Modern Coking Practice, Including Analysis of Materials and Products.

By J. E. Christopher and T. H. Byrom. This, the standard work on the subject, has just been revised and is now issued in two volumes. It is a practical work for those engaged in Coke manufacture and the recovery of By-products. Fully illustrated with folding plates. It has been the aim of the authors, in preparing this book, to produce one which shall be of use and benefit to those who are associated with, or interested in, the modern developments of the industry. Among the chapters contained in Volume I are: Introduction; Classification of Fuels; Impurities of Coals; Coal Washing; Sampling and Valuation of Coals, etc.; Chlorific Power of Fuels; History of Coke Manufacture; Developments in Coke Oven Design; Recent Types of Coke Ovens; Mechanical Appliances at Coke Ovens; Chemical and Physical Examination of Coke. Volume II covers By-products. Each volume is fully illustrated, with folding plates.

Price, $3.00 per volume

COMPRESSED AIR

Compressed Air in all Its Applications.

By Gardner D. Hiscox. This is the most complete book on the subject of air that has ever been issued, and its thirty-five chapters include about every phase of the subject one can think of. It may be called an encyclopedia of compressed air. It is written by an expert, who, in its 665 pages, has dealt with the subject in a comprehensive manner, no phase of it being omitted. Over 500 illustrations. Fifth Edition, revised and enlarged. Cloth bound, **$6.00.** Half Morocco, revised and enlarged. Cloth bound. **Price, $6.00**

CONCRETE

Concrete Wall Forms.

By A. A. Houghton. A new automatic wall clamp, is illustrated with working drawings. Other types of wall forms, clamps, separators, etc., are also illustrated and explained.

Price, 75 cents

Concrete Floors and Sidewalks.

By A. A. HOUGHTON. The molds for molding squares, hexagonal and many other styles of mosaic floor and sidewalk blocks are fully illustrated and explained. **Price, 75 cents**

Practical Concrete Silo Construction.

By A. A. HOUGHTON. Complete working drawings and specifications are given for several styles of concrete silos, with illustrations of molds for monolithic and block silos. The tables, data, and information presented in this book are of the utmost value in planning and constructing all forms of concrete silos. **Price, 75 cents**

Molding Concrete Bath Tubs, Aquariums and Natatoriums.

By A. A. HOUGHTON. Simple molds and instruction are given for molding different styles of concrete bath tubs, swimming pools, etc. **Price, 75 cents**

Molding Concrete Chimneys, Slate and Roof Tiles.

By A. A. HOUGHTON. The manufacture of all types of concrete slate and roof tile is fully treated. Valuable data on all forms of reinforced concrete roofs are contained within its pages. The construction of concrete chimneys by block and monolithic systems is fully illustrated and described. A number of ornamental designs of chimney construction with molds are shown in this valuable treatise. **75 cents**

Molding and Curing Ornamental Concrete.

By A. A. HOUGHTON. The proper proportions of cement and aggregates for various finishes, also the methods of thoroughly mixing and placing in the molds, are fully treated. An exhaustive treatise on this subject that every concrete worker will find of daily use and value. **Price, 75 cents**

Concrete Monuments, Mausoleums and Burial Vaults.

By A. A. HOUGHTON. The molding of concrete monuments to imitate the most expensive cut stone is explained in this treatise, with working drawings of easily built molds. Cutting inscriptions and designs is also fully treated. **75 cents**

Concrete Bridges, Culverts and Sewers.

By A. A. HOUGHTON. A number of ornamental concrete bridges with illustrations of molds are given. A collapsible center of core for bridges, culverts and sewers is fully illustrated with detailed instructions for building. **75 cents**

Constructing Concrete Porches.

By A. A. HOUGHTON. A number of designs with working drawings of molds are fully explained so any one can easily construct different styles of ornamental concrete porches without the purchase of expensive molds. **Price, 75 cents**

Molding Concrete Flower Pots, Boxes, Jardinieres, Etc.

By A. A. Houghton. The molds for producing many original designs of flower pots, urns, flower boxes, jardinieres, etc., are fully illustrated and explained, so the worker can easily construct and operate same. **Price, 75 cents**

Molding Concrete Fountains and Lawn Ornaments.

By A. A. Houghton. The molding of a number of designs of lawn seats, curbing, hitching posts, pergolas, sun dials and other forms of ornamental concrete, for the ornamentation of lawns and gardens, is fully illustrated and described. **75c.**

Concrete on the Farm and in the Shop.

By H. Colvin Campbell. This is a new book from cover to cover, illustrating and describing in plain, simple language many of the numerous appliances of concrete within the range of the home worker. Among the subjects treated are: Principles of reinforcing; methods of protecting concrete so as to insure proper hardening; home-made mixers; mixing by hand and machine; form construction, described and illustrated by drawings and photographs; construction of concrete walls and fences; concrete fence posts; concrete gate posts; corner posts; clothes line posts; grape arbor posts; tanks; troughs; cisterns: hog wallows; feeding floors and barnyard pavements; foundations; well curbs and platforms; indoor floors; sidewalks; steps; concrete hotbeds and cold frames; concrete slab roofs; walls for buildings; repairing leaks in tanks and cisterns; and all topics associated with these subjects as bearing upon securing the best results from concrete are dwelt upon at sufficient length in plain every-day English so that the inexperienced person desiring to undertake a piece of concrete construction can, by following the directions set forth in this book, secure 100 per cent success every time. A number of convenient and practical tables for estimating quantities, and some practical examples, are also given. 150 pages, 51 illustrations. **Price, $1.00**

Concrete From Sand Molds.

By A. A. Houghton. A practical work treating on a process which has heretofore been held as a trade secret by the few who possessed it, and which will successfully mold every and any class of ornamental concrete work. The process of molding concrete with sand molds is of the utmost practical value, possessing the manifold advantages of a low cost of molds, the ease and rapidity of operation, perfect details to all ornamental designs, density and increased strength of the concrete, perfect curing of the work without attention and the easy removal of the molds regardless of any undercutting the design may have. 192 pages. Fully illustrated. Cloth. **Price, $2.00**

Ornamental Concrete Without Molds.

By A. A. Houghton. The process for making ornamental concrete without molds has long been held as a secret, and now, for the first time, this process is given to the public. The book reveals the secret and is the only book published which explains a simple, practical method whereby the concrete worker is enabled, by employing wood and metal templates of different designs, to mold or model in concrete any cornice, archivolt, column, pedestal, base cap, urn or pier in a monolithic form—right upon the job. These may be molded in units or blocks, and then built up to suit the specifications demanded. This work is fully illustrated, with detailed engravings. Cloth. **Price, $2.00**

Popular Handbook for Cement and Concrete Users.

By Myron H. Lewis. Everything of value to the concrete user is contained, including kinds of cement employed in construction, concrete architecture, inspection and testing, waterproofing, coloring and painting, rules tables, working and cost data. The book comprises thirty-three chapters. A valuable addition to the library of every cement and concrete user. Cloth, 430 pages, 126 illustrations. **Price, $3.00**

Waterproofing Concrete.

By Myron H. Lewis. Modern methods of waterproofing concrete and other structures. A condensed statement of the principles, rules and precautions to be observed in waterproofing and damp-proofing structures and structural materials. Paper binding. Illustrated. Second Edition. **75 cents**

DIES—METAL WORK

Dies; Their Construction and Use for the Modern Working of Sheet Metals.

By J. V. Woodworth. A new book by a practical man, for those who wish to know the latest practice in the working of sheet metals. It shows how dies are designed, made and used, and those who are engaged in this line of work can secure many valuable suggestions. Sixth revised edition. 525 illustrations, 394 pages. Cloth. **Price, $3.50**

Punches, Dies and Tools for Manufacturing in Presses.

By J. V. Woodworth. An encyclopedia of die-making, punch-making, die-sinking, sheet-metal working, and making of special tools, subpresses, devices and mechanical combinations for punching, cutting, bending, forming, piercing, drawing, compressing, and assembling sheet-metal parts and also articles of other materials in machine tools. This is a distinct work from the author's book entitled "Dies; Their Construction and Use." 500 pages, 700 engravings. Second edition. Cloth. **Price, $4.50**

Drop Forging, Die-Sinking and Machine-Forming of Steel.

By J. V. Woodworth. The processes of die-sinking and force-making, which are thoroughly described and illustrated in this admirable work, are rarely to be found explained in such a clear and concise manner as is here set forth. The process of die-sinking relates to the engraving or sinking of the female or lower dies, such as are used for drop forgings, hot and cold machine forging, swedging and the press working of metals. The process of force-making relates to the engraving or raising of the male or upper dies used in producing the lower dies for the press-forming and machine-forging of duplicate parts of metal. The book contains eleven chapters, and the information contained in these chapters is just what will prove most valuable to the forged-metal worker. 304 detailed illustrations. 341 pages, cloth. **Price, $3.00**

DICTIONARIES

Aviation Terms—English-French; French-English.

A complete glossary of practically all terms used in aviation, having lists in both French and English with equivalents in either language. A very valuable book compiled by Lieuts. Victor W. Pagé and Paul Montariol. **Price, $1.00**

Standard Electrical Dictionary.

By PROF. T. O'CONOR SLOANE. Just issued an entirely new edition brought up to date and greatly enlarged—as a reference book this work is beyond comparison as it contains over 700 pages, nearly 500 illustrations, and definitions of about 6,000 distinct words, terms and phrases. The definitions are terse and concise and includes every term used in electrical science.

In its arrangement and typography the book is very convenient. The word or term defined is printed in black faced type which readily catches the eye, while the body of the page is in smaller but distinct type. The definitions are well worded, and so as to be understood by the non-technical reader. The general plan is to give an exact, concise definition, and then amplify and explain in a more popular way. Synonyms are also given, and references to other words and phrases are made. This work is absolutely indispensable to all in any way interested in electrical science, from the higher electrical expert to the everyday electrical workman. In fact, it should be in the possession of all who desire to keep abreast with the progress of this branch of science. 1920 enlarged edition. Nearly 800 pages and nearly 400 illustrations. **Price, $5.00**

DRAWING—SKETCHING PAPER

Linear Perspective Self-Taught.

By HERMAN T. C. KRAUS. This work gives the theory and practice of linear perspective, as used in architectural, engineering and mechanical drawings. The arrangement of the book is good; the plate is on the left-hand, while the descriptive text follows on the opposite page, so as to be readily referred to. A self-explanatory linear perspective chart is included in the second revised edition. Cloth. **Price, $3.00**

Self-Taught Mechanical Drawing and Elementary Machine Design.

By F. L. Sylvester, M.E., Draftsman, with additions by Erik Oberg, associate editor of "Machinery." A practical elementary treatise on Mechanical Drawing and Machine Design, comprising the first principles of geometric and mechanical drawing, workshop mathematics, mechanics, strength of materials and the calculation and design of machine details, compiled for the use of practical mechanics and young draftsmen. 330 pages, 215 engravings, cloth. **Price, $2.50**

A New Sketching Paper.

A new specially ruled paper to enable you to make sketches or drawings in isometric perspective without any figuring or fussing. It is being used for shop details as well as for assembly drawings, as it makes one sketch do the work of three, and no workman can help seeing just what is wanted. Pads of 40 sheets, 6 x 9 inches, **Price, 25c.;** 9 x 12 inches, **Price, 50c.;** 12 x 18 inches, **Price, $1.00.**

Practical Perspective.

By Richards and Colvin. Shows just how to make all kinds of mechanical drawings in the only practical perspective isometric. Makes everything plain so that any mechanic can understand a sketch or drawing in this way. Saves time in the drawing room and mistakes in the shops. Contains practical examples of various classes of work. Third edition. Limp cloth. **Price, 75 cents**

ELECTRICITY

Arithmetic of Electricity.

By Prof. T. O'Conor Sloane. A practical treatise on electrical calculations of all kinds reduced to a series of rules, all of the simplest forms, and involving only ordinary arithmetic; each rule illustrated by one or more practical problems with detailed solution of each one. This book is classed among the most useful works published on the science of electricity, covering as it does the mathematics of electricity in a manner that will attract the attention of those who are not familiar with algebraical formulas. 200 pages. 1920 Revised and Enlarged edition. **Price, $1.50**

Dynamo Building for Amateurs, or How to Construct a Fifty Watt Dynamo.

By Arthur J. Weed. A practical treatise showing in detail the construction of a small dynamo or motor, the entire machine work of which can be done on a small foot lathe. Dimensioned working drawings are given for each piece of machine work, and each operation is clearly described. This machine, when used as a dynamo, has an output of fifty watts; when used as a motor it will drive a small drill press or lathe. It can be used to drive a sewing machine on any and all ordinary work. The book is illustrated with more than sixty original engravings showing the actual construction of the different parts. **Price, Cloth, $1.00**

Electric Bells.

By M. B. Sleeper. A complete treatise for the practical worker in installing, operating and testing bell circuits, burglar alarms, thermostats and other apparatus used with electric bells. Both the electrician and the experimenter will find in this book new material which is essential in their work. Tools, bells, batteries, unusual circuits, burglar alarms, annunciators, systems, thermostats, circuit breakers, time alarms, and other apparatus used in bell circuits are described from the standpoints of their application, construction, and repair. The detailed instructions for building the apparatus will appeal to the experimenter particularly. The practical worker will find the chapters on Wiring Calculation of Wire Sizes and Magnet Windings, Upkeep of Systems and the Location of Faults of the greatest value in their work. 124 pages. Fully illustrated. **Price, 75 cents.**

Commutator Construction

By Wm. Baxter, Jr. The business end of dynamo or motor of the direct current type is the commutator. This book goes into the designing, building and maintenance of commutators, shows how to locate troubles and how to remedy them; everyone who fusses with dynamos needs this. Fourth edition. **Price, 35 cents.**

Dynamos and Electric Motors and All About Them.

By Edward Trevert. This volume gives practical directions for building a two H. P. Dynamo of the Edison type capable of lighting about fifty mazda lamps of the 20 watt size. In addition, it gives directions for building two small electric motors suitable for running sewing machines. The concluding chapter describes the construction of a simple bichromate battery adapted for running electric motors. 96 pages. Fully illustrated with detail drawings. Cloth. **Price, $1.00.**

Construction of a Transatlantic Wireless Receiving Set.

By L. G. Pacent and T. S. Curtis. A work for the Radio student who desires to construct and operate apparatus that will permit of the reception of messages from the large stations in Europe with an aerial of amateur proportions. 36 pages. 23 illustrations, cloth. **Price, 35 cents.**

Electric Toy Making, Dynamo Building and Electric Motor Construction.

This work treats of the making at home of electrical toys, electrical apparatus, motors, dynamos and instruments in general and is designed to bring within the reach of young and old the manufacture of genuine and useful electrical appliances. 210 pages, cloth. Fully illustrated. Twentieth edition, enlarged. **Price, $1.50**

Experimental High Frequency Apparatus, How to Make and Use It.

By Thomas Stanley Curtis. 69 pages, illustrated. **Price, 50 cents.**

Electrician's Handy Book.

By PROF. T. O'CONOR SLOANE. This work has just been revised and much enlarged. It is intended for the practical electrician who has to make things go. The entire field of electricity is covered within its pages. It is a work of the most modern practice, written in a clear, comprehensive manner, and covers the subject thoroughly, beginning at the A B C of the subject, and gradually takes you to the more advanced branches of the science. It teaches you just what you should know about electricity. A practical work for the practical man. Contains forty-eight chapters.

The publishers consider themselves fortunate in having secured the services of such a well and favorably known writer as Prof. Sloane, who has with the greatest care completed a master work in concise form on this all important subject. 600 engravings, 840 pages, handsomely bound in cloth. 1920 Edition. **Price, $4.00**

Electricity Simplified.

By PROF. T. O'CONOR SLOANE. The object of 'Electricity Simplified" is to make the subject as plain as possible and to show what the modern conception of electricity is; to show how two plates of different metals immersed in acid can send a message around the globe; to explain how a bundle of copper wire rotated by a steam engine can be the agent in lighting our streets, to tell what the volt, ohm and ampere are, and what high and low tension mean; and to answer the questions that perpetually arise in the mind in this age of electricity. 172 pages. Illustrated. Thirteenth edition. Cloth. **Price, $1.50**

Electric Wiring, Diagrams and Switchboards.

By NEWTON HARRISON, with additions by THOMAS POPPE. This is the only complete work issued showing and telling you what you should know about direct and alternating current wiring. The work is free from advanced technicalities and mathematics, arithmetic being used throughout. It is in every respect a handy, well-written, instructive, comprehensive volume on wiring for the wireman, foreman, contractor or electrician. Third revised edition. 303 pages, 130 illustrations. Cloth. **Price, $2.50**

House Wiring.

By THOMAS W. POPPE. Describing and illustrating up-to-date methods of installing electric light wiring. Contains just the information needed for successful wiring of a building. Fully illustrated with diagrams and plans. It solves all wiring problems and contains nothing that conflicts with the rulings of the National Board of Fire Underwriters. Third edition revised and enlarged. 125 pages, fully illustrated, flexible cloth. **Price, 75 cents**

High Frequency Apparatus, Its Construction and Practical Application.

By THOMAS STANLEY CURTIS. The most comprehensive and thorough work on this interesting subject ever produced. The book is essentially practical in its treatment and it constitutes an accurate record of the researches of its author over a period of several years, during which time dozens of coils were built and experimented with. New revised and enlarged 1920 edition. 275 pages. **Price, $3.00**

How to Become a Successful Electrician.

By PROF. T. O'CONOR SLOANE. An interesting book from cover to cover. Telling in simplest language the surest and easiest way to become a successful electrician. The studies to be followed, methods of work, field of operation and the requirements of the successful electrician are pointed out and fully explained. 202 pages. Illustrated. Eighteenth revised edition. Cloth. **Price, $1.50**

Standard Electrical Dictionary.

By PROF. T. O'CONOR SLOANE. Just issued an entirely new edition brought up to date and greatly enlarged—as a reference book this work is beyond comparison as it contains over 700 pages, nearly 500 illustrations, and definitions of about 6,000 distinct words, terms and phrases. The definitions are terse and concise and includes every term used in electrical science.

In its arrangement and typography the book is very convenient. The word or term defined is printed in black faced type which readily catches the eye, while the body of the page is in smaller but distinct type. The definitions are well worded, and so as to be understood by the non-technical reader. The general plan is to give an exact, concise definition, and then amplify and explain in a more popular way. Synonyms are also given, and references to other words and phrases are made. This work is absolutely indispensable to all in any way interested in electrical science, from the higher electrical expert to the everyday electrical workman. In fact, it should be in the possession of all who desire to keep abreast with the progress of this branch of science. 1920 enlarged edition. Nearly 800 pages. 400 illustrations. **Price, $5.00**

Storage Batteries Simplified.

By VICTOR W. PAGÉ. M.S.A.E. This is the most thorough and authoritative treatise ever published on this subject. It is written in easily understandable, non-technical language so that any one may grasp the basic principles of storage battery action as well as their practical industrial applications. All electric and gasoline automobiles use storage batteries. Every automobile repairman, dealer or salesman should have a good knowledge of maintenance and repair of these important elements of the motor car mechanism. This book not only tells how to charge, care for and rebuild storage batteries but also outlines all the industrial uses. Learn how they run street cars, locomotives and factory trucks. Get an understanding of the important functions they perform in submarine boats, isolated lighting plants, railway switch and signal systems, marine applications, etc. This book tells how they are used in central station standby service, for starting automobile motors and in ignition systems. Every practical use of the modern storage battery is outlined in this treatise. 208 pages, fully illustrated. **Price, $2.00**

Wiring a House.

By HERBERT PRATT. Shows a house already built; tells just how to start about wiring it; where to begin; what wire to use; how to run it according to insurance rules; in fact, just the information you need. Directions apply equally to a shop. Fourth edition. **Price, 35 cents**

Switchboards.

By WILLIAM BAXTER, JR. This book appeals to every engineer and electrician who wants to know the practical side of things. All sorts and conditions of dynamos, connections and circuits are shown by diagram and illustrate just how the switchboard should be connected. Includes direct and alternating current boards, also those for arc lighting, incandescent and power circuits. Special treatment on high voltage boards for power transmission. Second edition. 190 pages. Illustrated. **Price, $2.00**

Telephone Construction, Installation, Wiring, Operation and Maintenance.

By W. H. RADCLIFFE and H. C. CUSHING. This book gives the principles of construction and operation of both the Bell and Independent instruments; approved methods of installing and wiring them; the means of protecting them from lightning and abnormal currents; their connection together for operation as series or bridging stations; and rules for their inspection and maintenance. Line wiring and the wiring and operation of special telephone systems are also treated. 224 pages, 132 illustrations. Second revised edition. **Price, $1.50**

Wireless Telegraphy and Telephony Simply Explained.

By ALFRED P. MORGAN. This is undoubtedly one of the most complete and comprehensive treatises on the subject ever published, and a close study of its pages will enable one to master all the details of the wireless transmission of messages. The author has filled a long-felt want and has succeeded in furnishing a lucid, comprehensible explanation in simple language of the theory and practise of wireless telegraphy and telephony. Third edition. 154 pages, 156 engravings. **Price, $1.50**

Radio Time Signal Receiver.

By AUSTIN C. LESCARBOURA. This new book, "A Radio Time Signal Receiver," tells you how to build a simple outfit designed expressly for the beginner. You can build the outfits in your own workshop and install them for jewelers either on a one-payment or a rental basis. The apparatus is of such simple design that it may be made by the average amateur mechanic possessing a few ordinary tools. 42 pages. Paper. **Price, 35 cents**

Experimental Wireless Stations.

By P. E. EDELMAN. The theory, design, construction and operation is fully treated including Wireless Telephony, Vacuum Tube, and quenched spark systems. The new enlarged 1920 edition is just issued and is strictly up to date, correct and complete. This book tells how to make apparatus to not only hear all telephoned radio messages, but also how to make simple equipment that works for transmission over reasonably long distances. Then there is a host of new information included. The first and only book to give you all the recent important radio improvements, some of which have never before been published. 24 chapters. 167 illustrations. **Price, $2.50**

FUEL

Combustion of Coal and the Prevention of Smoke.

By WM. M. BARR. This book has been prepared with special reference to the generation of heat by the combustion of the common fuels found in the United States, and deals particularly with the conditions necessary to the economic and smokeless combustion of bituminous coals in stationary and locomotive steam boilers. The presentation of this important subject is systematic and progressive. The arrangement of the book is in a series of practical questions, to which are appended accurate answers, which describe in language, free from technicalities, the several processes involved in the furnace combustion of American fuels; it clearly states the essential requisites for perfect combustion, and points out the best methods for furnace construction for obtaining the greatest quantity of heat from any given quality of coal. Nearly 350 pages, fully illustrated. Fifth edition.

Price, $1.50

Smoke Prevention and Fuel Economy.

By BOOTH and KERSHAW. As the title indicates, this book of 197 pages and 75 illustrations deals with the problem of complete combustion, which it treats from the chemical and mechanical standpoints, besides pointing out the economical and humanitarian aspects of the question. **Price, $3.00**

GAS ENGINES AND GAS

Gas, Gasoline and Oil Engines.

By GARDNER D. HISCOX. Revised by VICTOR W. PAGÉ. Just issued new, revised and enlarged edition. Every user of a gas engine needs this book. Simple, instructive and right up-to-date. The only complete work on the subject. Tells all about internal combustion engineering, treating exhaustively on the design, construction and practical application of all forms of gas, gasoline, kerosene and crude petroleum-oil engines. Describes minutely all auxiliary systems, such as lubrication, carburetion and ignition. Considers the theory and management of all forms of explosive motors for stationary and marine work, automobiles, aeroplanes and motorcycles. Includes also Producer Gas and Its Production. Invaluable instructions for all students, gas-engine owners, gas-engineers, patent experts, designers, mechanics, draftsmen and all having to do with the modern power. Illustrated by over 400 engravings, many specially made from engineering drawings, all in correct proportion. Nearly 700 octavo pages and 500 engravings. **Price, net, $3.00**

Gasoline Engines: Their Operation, Use and Care.

By A. HYATT VERRILL. A comprehensive, simple and practical work, treating of gasoline engines for stationary, marine or vehicle use; their construction, design, management, care, operation, repair, installation and troubles. A complete glossary of technical terms and an alphabetically arranged table of troubles and symptoms form a most valuable and unique feature of the book. 5¼ x 7½. Cloth. 275 pages, 152 illustrations.

Price, $2.00

Gas Engine Construction.

Or How to Build a Half-Horse-power Gas Engine. By PARSELL and WEED. A practical treatise describing the theory and principles of the action of gas engines of various types, and the design and construction of a half-horse-power gas engine, with illustrations of the work in actual progress, together with dimensioned working drawings giving clearly the sizes of the various details. 300 pages. Third edition. Cloth. **Price, $3.00**

Chemistry of Gas Manufacture.

By H. M. ROYLES. This book covers points likely to arise in the ordinary course of the duties of the engineer or manager of a gas works not large enough to necessitate the employment of a separate chemical staff. It treats of the testing of the raw materials employed in the manufacture of illuminating coal gas and of the gas produced. The preparation of standard solutions is given as well as the chemical and physical examination of gas coal. 5¾ x 8¾. Cloth. 328 pages, 82 illustrations, 1 colored plate. **Price, $5.00**

Modern Gas Engines and Producer Gas Plants.

By R. E. MATHOT, M.E. A practical treatise of 320 pages, fully illustrated by 175 detailed illustrations, setting forth the principles of gas engines and producer design, the selection and installation of an engine, conditions of perfect operation, producer-gas engines and their possibilities, the care of gas engines and producer-gas plants, with a chapter on volatile hydrocarbon and oil engines. This book has been endorsed by Dugal Clerk as a most useful work for all interested in gas engine installation and producer gas. **$3.00**

The Gasoline Engine on the Farm: Its Operation, Repair and Uses.

By XENO W. PUTNAM. A useful and practical treatise on the modern gasoline and kerosene engine, its construction, management, repair and the many uses to which it can be applied in present-day farm life. It considers all the various household, shop and field uses of this up-to-date motor and includes chapters on engine installation, power transmission and the best arrangement of the power plant in reference to the work. 5¼ x 7½. Cloth. 527 pages, 179 illustrations. **Price, $3.00**

How to Run and Install Two- and Four-Cycle Marine, Gasoline Engines.

By C. VON CULIN. New revised and enlarged edition just issued. The object of this little book is to furnish a pocket instructor for the beginner, the busy man who uses an engine for pleasure or profit, but who does not have the time or inclination for a technical book, but simply to thoroughly understand how to properly operate, install and care for his own engine. The index refers to each trouble, remedy and subject alphabetically. Being a quick reference to find the cause, remedy and prevention for troubles, and to become an expert with his own engine. Pocket size. Paper binding. **Price, 25 cents**

Modern Gas Tractor, Its Construction, Utility, Operation and Repair.

By VICTOR W. PAGÉ. Treats exhaustively on the design and construction of farm tractors and tractor power-plants, and gives complete instructions on their care, operation and repair. All types and sizes of gasoline, kerosene and oil tractory are described, and every phase of traction engineering practice fully covered. Invaluable to all desiring reliable information on gas motor propelled traction engines and their use. Second edition revised by much additional matter. 5¼ x 7½. Cloth, 504 pages, 228 illustrations. 3 folding plates. **Price, $3.00**

GEARING AND CAMS

Bevel Gear Tables.

By D. AG. ENGSTROM. No one who has to do with bevel gears in any way should be without this book. The designer and draftsman will find it a great convenience, while to the machinist who turns up the blanks or cuts the teeth it is invaluable, as all needed dimensions are given and no fancy figuring need be done. Third edition. Cloth. **$1.50**

Change Gear Devices.

By OSCAR E. PERRIGO. A book for every designer, draftsman and mechanic who is interested in feed changes for any kind of machines. This shows what has been done and how. Gives plans, patents and all information that you need. Saves hunting through patent records and reinventing old ideas. A standard work of reference. Third edition. **$1.50**

Drafting of Cams.

By LOUIS ROUILLION. The laying out of cams is a serious problem unless you know how to go at it right. This puts you on the right road for practically any kind of cam you are likely to run up against. Third edition. **35 cents**

HYDRAULICS

Hydraulic Engineering.

By GARDNER D. HISCOX. A treatise on the properties, power, and resources of water for all purposes. Including the measurements of streams; the flow of water in pipes or conduits; the horse-power of falling water; turbine and impact water-wheels; wave-motors, centrifugal, reciprocating and air-lift pumps. With 300 figures and diagrams and 36 practical tables. 320 pages. **Price, $4.50**

ICE AND REFRIGERATION

Pocketbook of Refrigeration and Ice Making.

By A. J. WALLIS-TAYLOR. This is one of the latest and most comprehensive reference books published on the subject of refrigeration and cold storage. It explains the properties and refrigerating effect of the different fluids in use, the management of refrigerating machinery and the contsruction and insulation of cold rooms with their required pipe surface for different degrees of cold; freezing mixtures and non-freezing brines, temperatures of cold rooms for all kinds of provisions, cold storage charges for all classes of goods, ice making and storage of ice, data and memoranda for constant reference by refrigerating engineers, with nearly one hundred tables containing valuable references to every fact and condition required in the installment and operation of a refrigerating plant. New edition just published. **Price, $2.00**

INVENTIONS—PATENTS

Inventor's Manual, How to Make a Patent Pay.

This is a book designed as a guide to inventors in perfecting their inventions, taking out their patents, and disposing of them. It is not in any sense a Patent Solicitor's circular nor a Patent Broker's advertisement. No advertisements of any description appear in the work. It is a book containing a quarter of a century's experience of a successful inventor, together with notes based upon the experience of many other inventors. Revised and enlarged second edition. Nearly 150 pages. Illustrated. **Price $1.25**

KNOTS

Knots, Splices and Rope Work.

By A. Hyatt Verrill. This is a practical book giving complete and simple directions for making all the most useful and ornamental knots in common use, with chapters on Splicing, Pointing, Seizing, Serving, etc. This book is fully illustrated with 154 original engravings, which show how each knot, tie or splice is formed, and its appearance when finished. The book will be found of the greatest value to campers, yachtsmen, travelers or Boy Scouts, in fact, to anyone having occasion to use or handle rope or knots for any purpose. The book is thoroughly reliable and practical, and is not only a guide but a teacher. It is the standard work on the subject. Second edition revised. 128 pages, 154 original engravings. **Price, $1.00**

LATHE WORK

Complete Practical Machinist.

By Joshua Rose. The new, twentieth revised and enlarged edition is now ready. This is one of the best-known books on machine-shop work, and written for the practical workman in the language of the workshop. It gives full, practical intsrsuctions on the use of all kinds of metal-working tools, both hand and machine, and tells how the work should be properly done. It covers lathe work, vise work, drills and drilling, taps and dies, hardening and tempering, the making and use of tools, tool grinding, marking out work, machine tools, etc. No machinist's library is complete without this volume. 547 pages, 432 illustrations. (1920.) **Price $3.00**

The Lathe—Its Design, Construction and Operation, With Practical Examples of Lathe Work.

By OSCAR E. PERRIGO. A new revised edition, and the only complete American work on the subject, written by a man who knows not only how work ought to be done, but who also knows how to do it, and how to convey this knowledge to others. It is strictly up-to-date in its descriptions and illustrations. Lathe history and the relations of the lathe to manufacturing are given; also a description of the various devices for feeds and thread cutting mechanisms from early efforts in this direction to the present time. Lathe design is thoroughly discussed, including back gearing, driving cones, thread-cutting gears, and all the essential element of the modern lathe. The classification of lathes is taken up, giving the essential differences of the several types of lathes including, as is usually understood, engine lathes, bench lathes, speed lathes, forge lathes, gap lathes, pulley lathes, forming lathes, multiple-spindle lathes, rapid-reduction lathes, precision lathes, turret lathes, special lathes, electrically-driven lathes, etc. In addition to the complete exposition on construction and design, much practical matter on lathe installation, care and operation has been incorporated in the enlarged new edition. All kinds of lathe attachments for drilling, milling, etc., are described and complete instructions are given to enable the novice machinist to grasp the art of lathe operation as well as the principles involved in design. A number of difficult machining operations are described at length and illustrated. The new edition has nearly 500 pages and 350 illustrations. **Price, $3.00**

Turning and Boring Tapers.

By FRED H. COLVIN. There are two ways to turn tapers; the right way and one other. This treatise has to do with the right way; it tells you how to start the work properly, how to set the lathe, what tools to use and how to use them, and forty and one other little things that you should follow. Fourth edition. **Price, 35 cents**

LIQUID AIR

Liquid Air and the Liquefaction of Gases.

By T. O'CONOR SLOANE. The third revised edition of this book has just been issued. Much new material is added to it; and the all important uses of liquid air and gas processes in modern industry, in the production especially of nitrogen compounds, are described. The book gives the history of the theory, discovery, and manufacture of Liquid Air, and contains an illustrated description of all the experiments that have excited the wonder of audiences all over the country. It shows how liquid air, like water, is carried hundreds of miles and is handled in open buckets. It tells what may be expected from it in the near future. A book that renders simple one of the most perplexng chemical problems of the century. Startling developments illustrated by actual experiments. It is not only a work of scientific interest and authority, but is intended for the general reader, being written in a popular style—easily understood by everyone. 400 pages fully illustrated. (1920.) **Price, $3.00**

Air-Brake Catechism.

By ROBERT H. BLACKALL. This book is a standard text book. It is the only practical and complete work published. Treats on the equipment manufactured by the Westinghouse Air Brake Company, including the E-T Locomotive Brake Equipment, the K (Quick-Service) Triple Valve for freight service; the L High Speed Triple Valve; the P-C Passenger Brake Equipment, and the Cross Compound Pump. The operation of all parts of the apparatus is explained in detail and a practical way of locating their peculiarities and remedying their defects is given. Endorsed and used by air-brake instructors and examiners on nearly every railroad in the United States. Twenty-sixth edition. 411 pages, fully illustrated with folding plates and diagrams. New edition.
Price, $2.50

Application of Highly Superheated Steam to Locomotives.

By ROBERT GARBE. A practical book which cannot be recommended too highly to those motive-power men who are anxious to maintain the highest efficiency in their locomotives. Contains special chapters on Generation of Highly Superheated Steam; Superheated Steam and the Two-Cylinder Simple Engine; Compounding and Superheating; Designs of Locomotive Superheaters; Constructive Details of Locomotives Using Highly Superheated Steam. Experimental and Working Results. Illustrated with folding plates and tables. Cloth. **Price, $3.00**

Combustion of Coal and the Prevention of Smoke.

By WM. M. BARR. To be a success a fireman must be "Light on Coal." He must keep his fire in good condition, and prevent, as far as possible, the smoke nuisance. To do this, he should know how coal burns, how smoke is formed and the proper burning of fuel to obtain the best results. He can learn this, and more too, from Barr's "Combustion of Coal." It is an absolute authority on all questions relating to the firing of a locomotive. Fifth edition. Nearly 350 pages, fully illustrated. **Price, $1.50**

Diary of a Round-House Foreman.

By T. S. REILLY. This is the greatest book of railroad experiences ever published. Containing a fund of information and suggestions along the line of handling men, organizing, etc., that one cannot afford to miss. 176 pages. **Price, $1.25**

Link Motions, Valves and Valve Setting.

By FRED H. COLVIN, Associate Editor of "American Machinist." A handy book that clears up the mysteries of valve setting. Shows the different valve gears in use, how they work, and why. Piston and slide valves of different types are illustrated and explained. A book that every railroad man in the motive-power department ought to have. Fully illustrated. New revised and enlarged edition just published.
Price, 75 cents

Train Rule Examinations Made Easy.

By G. E. Collingwood. This is the only practical work on train rules in print. Every detail is covered, and puzzling points are explained in simple, comprehensive language, making it a practical treatise for the train dispatcher, engineman, trainman and all others who have to do with the movements of trains. Contains complete and reliable information of the Standard Code of Train Rules for single track. Shows signals in colors, as used on the different roads. Explains fully the practical application of train orders, giving a clear and definite understanding of all orders which may be used. Second edition revised. 256 pages. Fully illustrated with train signals in colors. **Price, $1.50**

Locomotive Boiler Construction.

By Frank A. Kleinhans. The only book showing how locomotive boilers are built in modern shops. Shows all types of boilers used; gives details of construction; practical facts, such as life of riveting punches and dies, work done per day, allowance for bending and flanging sheets and other data that means dollars to any railroad man. Second edition. 451 pages, 334 illustrations. Six folding plates. Cloth. **Price, $3.50**

Locomotive Breakdowns and Their Remedies.

By Geo. L. Fowler. Revised by Wm. W. Wood, Air-Brake Instructor. Pocket edition. It is out of the question to try and tell you about every subject that is covered in this pocket edition of Locomotive Breakdowns. Just imagine all the common troubles that an engineer may expect to happen some time, and then add all of the unexpected ones, troubles that could occur, but that you had never thought about, and you will find that they are all treated with the very best methods of repair. Walschaert Locomotive Valve Gear Troubles, Electric Headlight Troubles, as well as Questions and Answers on the Air Brake, are all included. Eighth edition. 294 pages. Fully illustrated. **Price, $1.50**

Locomotive Catechism.

By Robert Grimshaw. Twenty-eighth revised and enlarged edition. This may well be called an encyclopedia of the locomotive. Contains over 4,000 examination questions with their answers, including among them those asked at the first, second and third year's examinations. 825 pages, 437 illustrations and 3 folding plates. **Price, $2.50**

Westinghouse E. T. Air-Brake Instruction Pocketbook Catechism.

By Wm. W. Wood, Air-Brake Instructor. A practical work containing examination questions and answers on the E. T. Equipment. Covering what the E. T. Brake is. How it should be operated. What to do when defective. Not a question can be asked of the engineman up for promotion on either the No. 5 or the No. 6 E. T. equipment that is not asked and answered in the book. If you want to thoroughly understand the E. T. equipment get a copy of this book. It covers every detail. Makes air-brake troubles and examinations easy. Fully illustrated with colored plates, showing various pressures. Cloth. (1920.) **Price, $2.50**

Practical Instructor and Reference Book for Locomotive Firemen and Engineers.

By CHAS. F. LOCKHART. An entirely new book on the locomotive. It appeals to every railroad man, as it tells him how things are done and the right way to do them. Written by a man who has had years of practical experience in locomotive shops and on the road firing and running. The information given in this book cannot be found in any other similar treatise. Eight hundred and fifty-one questions with their answers are included, which will prove specially helpful to those preparing for examination. 368 pages, 88 illustrations. Cloth. **Price, $2.00**

Prevention of Railroad Accidents, or Safety in Railroading.

By GEORGE BRADSHAW. This book is a heart-to-heart talk with railroad employees, dealing with facts, not theories, and showing the men in the ranks, from every-day experience, how accidents occur and how they may be avoided. The book is illustrated with seventy original photographs and drawings showing the safe and unsafe methods of work. No visionary schemes, no ideal pictures. Just plain facts and practical suggestions are given. Every railroad employee who reads the book is a better and safer man to have in railroad service. It gives just the information which will be the means of preventing many injuries and deaths. All railroad employees should procure a copy; read it, and do their part in preventing accidents. 169 pages. Pocket size. Fully illustrated. **Price, 50 cents**

Walschaert Locomotive Valve Gear.

By WM. W. WOOD. If you would thoroughly understand the Walschaert Valve Gear, you should possess a copy of this book. The author divides the subject into four divisions, as follows: I. Analysis of the gear. II. Designing and erection of the gear. III. Advantages of the gear. IV. Questions and answers relating to the Walschaert Valve Gear. This book is specially valuable to those preparing for promotion. Third edition. 245 pages. Fully illustrated. Cloth. **Price, $2.50**

MACHINE SHOP PRACTICE

Modern Machine Shop Construction, Equipment and Management.

By OSCAR E. PERRIGO. The only work published that describes the Modern Machine Shop or Manufacturing Plant from the time the grass is growing on the site intended for it until the finished product is shipped. Just the book needed by those contemplating the erection of modern shop buildings, the rebuilding and reorganization of old ones or the introduction of Modern Shop Methods, Time and Cost Systems. It is a book written and illustrated by a practical shop man for practical shop men who are too busy to read theories and want facts. It is the most complete all-around book of its kind ever published. Second edition. 384 pages, 219 original and specially-made illustrations. **Price, $5.00**

Machine Shop Arithmetic.

By Colvin-Cheney. Most popular book for shop men. Shows how all shop problems are worked out and "why." Includes change gears for cutting any threads; drills, taps, shink and force fits; metric system of measurements and threads. Used by all classes of mechanics and for instruction in Y. M. C. A. and other schools. Seventh edition. 131 pages. **Price, 75 cents.**

Abrasives and Abrasive Wheels.

By Fred B. Jacobs. A new book for everyone interested in abrasives or grinding. A careful reading of the book will not only make mechanics better able to use abrasives intelligently, but it will also tell the shop superintendent of many short cuts and efficiency-increasing kinks. The economic advantage in using large grinding wheels are fully explained, together with many other things that will tend to give the superintendent or workman a keen insight into abrasive engineering. 340 pages, 200 illustrations. This is an indispensable book for every machinist. **Price, $3.00**

American Tool Making and Interchangeable Manufacturing.

By J. V. Woodworth. In its 500-odd pages the one subject only, Tool Making, and whatever relates thereto, is dealt with. The work stands without a rival. It is a complete practical treatise on the art of American Tool Making and system of interchangeable manufacturing as carried on to-day in the United States. In it are described and illustrated all of the different types and classes of small tools, fixtures, devices and special appliances which are in general use in all machine-manufacturing and metal-working establishments where economy, capacity and interchangeability in the production of machined metal parts are imperative. The science of jig making is exhaustively discussed, and particular attention is paid to drill jigs, boring, profiling and milling fixtures and other devices in which the parts to be machined are located and fastened within the contrivances. All of the tools, fixtures and devices illustrated and described have been or are used for the actual production of work, such as parts of drill presses, lathes, patented machinery, typewriters, electrical apparatus, mechanical appliances, brass goods, composition parts, mould products, sheet metal articles, drop forgings, jewelry, watches, medals, coins, etc. Second edition. 531 pages. **Price, $4.50**

Henley's Encyclopedia of Practical Engineering and Allied Trades.

Edited by Joseph G. Horner, A.M.I.Mech.E. This book covers the entire practice of Civil and Mechanical Engineering. The best known experts in all branches of engineering have contributed to these volumes. The Cyclopedia is admirably well adapted to the needs of the beginner and the self-taught practical man, as well as the mechanical engineer, designer, draftsman, shop superintendent, foreman and machinist.

It is a modern treatise in five volumes. Handsomely bound in half morocco, each volume containing nearly 500 pages, with thousands of illustrations, including diagrammatic and sectional drawings with full explanatory details.

Price, $30.00. For the complete set of five volumes.

"Shop Kinks."

By ROBERT GRIMSHAW. This shows special methods of doing work of various kinds, and releasing cost of production. Has hints and kinks from some of the largest shops in this country and Europe. You are almost sure to find some that apply to your work, and in such a way as to save time and trouble. 400 pages. Fifth edition. Cloth. **Price, $3.00**

Machine Shop Tools and Shop Practice.

By W. H. VANDERVOORT. A work of 555 pages and 673 illustrations, describing in every detail the construction, operation, and manipulation of both hand and machine tools. Includes chapters on filing, fitting, and scraping surfaces; on drills, reamers, taps, and dies; the lathe and its tools; planers, shapers, and their tools; milling machines and cutters; gear cutters and gear cutting; drilling machines and drill work; grinding machines and their work; hardening and tempering; gearing, belting, and transmission machinery; useful data and tables. Sixth edition. Cloth. **Price, $4.50**

Model Making.

By RAYMOND FRANCIS YATES. A new book for the mechanic and model maker. This is the first book of its kind to be published in this country and all those interested in model engineering should have a copy. The first eight chapters are devoted to such subjects as Silver Soldering, Heat Treatment of Steel, Lathe Work, Pattern Making, Grinding, etc. The remaining twenty-four chapters describe the construction of various models such as rapid fire naval guns, speed boats, model steam engines, turbines, etc.

This book must not be confused with those describing the construction of toys now on the market. It is a practical treatise on model engineering and construction. 400 pages. 301 illustrations. **Price, $3.00**

The Modern Machinist.

By JOHN T. USHER. This book might be called a compendium of shop methods, showing a variety of special tools and appliances which will give new ideas to many mechanics from the superintendent down to the man at the bench. It will be found a valuable addition to any machinist's library and should be consulted whenever a new or difficult job is to be done, whether it is boring, milling, turning, or planing, as they are all treated in a practical manner. Fifth edition. 320 pages, 250 illustrations. Cloth. **Price, $2.50**

Threads and Thread Cutting.

By COLVIN and STABEL. This clears up many of the mysteries of thread cutting, such as double and triple threads, internal threads, catching threads, use of hobs, etc. Contains a lot of useful hints and several tables. Third edition. **35 cents**

MARINE ENGINEERING

The Naval Architect's and Shipbuilder's Pocket-book

of Formulæ, Rules, and Tables and Marine Engineer's and Surveyor's Handy Book of Reference. By CLEMENT MACKROW and LLOYD WOOLLARD. The eleventh revised and enlarged edition of this most comprehensive work has just been issued. It is absolutely indispensable to all engaged in the Shipbuilding Industry, as it condenses into a compact form all data and formulæ that are ordinarily required. The book is completely up to date, including among other subjects a section on Aeronautics. 750 pages, limp leather binding.

Price, $6.00 net

Marine Engines and Boilers, Their Design and Construction. The Standard Book.

By DR. G. BAUER, LESLIE S. ROBERTSON and S. BRYAN DONKIN. In the words of Dr. Bauer, the present work owes its origin to an oft felt want of a condensed treatise embodying the theoretical and practical rules used in designing marine engines and boilers. The need of such a work has been felt by most engineers engaged in the construction and working of marine engines, not only by the younger men, but also by those of greater experience. The fact that the original German work was written by the chief engineer of the famous Vulcan Works, Stettin, is in itself a guarantee that this book is in all respects thoroughly up-to-date, and that it embodies all the information which is necessary for the design and construction of the highest types of marine engines and boilers. It may be said that the motive power which Dr. Bauer has placed in the fast German liners that have been turned out of late years from the Stettin Works represent the very best practice in marine engineering of the present day. The work is clearly written, thoroughly systematic, theoretically sound; while the character of the plans, drawings, tables, and statistics is without reproach. The illustrations are careful reproductions from actual working drawings, with some well-executed photographic views of completed engines and boilers. 744 pages, 550 illustrations, and numerous tables. Cloth. **Price, $10.00 net**

MANUAL TRAINING

Economics of Manual Training.

By LOUIS ROUILLION. The only book that gives just the information needed by all interested in manual training, regarding buildings, equipment and supplies. Shows exactly what is needed for all grades of the work from the Kindergarten to the High and Normal School. Gives itemized lists of everything needed and tells just what it ought to cost. Also shows where to buy supplies. Illustrated. Second edition. Cloth. **Price, $2.00**

MINING

Prospector's Field-Book and Guide.

By H. S. Osborn. 1920 edition, revised and enlarged by M. W. von Bernewitz. The last edition of this volume was published in 1910. It and the previous seven editions were suitable for those times. The new ninth (1920) edition will be found suitable for the present time. While the old-time prospector will always be an important factor, the knowledge of and search for the common and rarer minerals is bringing out men who are trained to some degree. In the field they need a handy and suggestive pocket-book containing hints on prospecting—where to search and how to test—couched in simple terms. The chapter on preliminary instructions covers the fundamentals of a study of the earth's crust. Then follow discussions on practical mineralogy, crystallography, the value of the blowpipe in prospecting, surveying, and chemical tests in the field. Separate chapters are given to the precious and base metals, also to the non-metallic minerals. The chapter on the non-ferrous or alloy group of minerals is entirely new, while the section on oil has been expanded. Surficial indications for copper receive full attention. The chapter on gems has been rewritten and matters concerning gemstones used for industrial purposes, such as abrasives, included. A general chapter covers many useful minerals and salts. An important guide and suggestive aid throughout the new book are the many brief descriptions of ore deposits of all minerals occurring in scattered parts of the world. No other prospector's book contains this class of information. In the appendix will be found numbers of useful tables, and a complete glossary of mining and mineralogical terms. The ninth edition of Osborn's "Prospector's Field Book and Guide" will be found up to date, worth while, and full value for the money asked. Flexible fabrikoid. 375 pages. 57 illustrations. 1920 edition. **Price, $3.00**

PATTERN MAKING

Practical Pattern Making.

By F. W. Barrows. This book, now in its second edition, is a comprehensive and entirely practical treatise on the subject of pattern making, illustrating pattern work in both wood and metal, and with definite instructions on the use of plaster of paris in the trade. It gives specific and detailed descriptions of the materials used by pattern makers and describes the tools; both those for the bench and the more interesting machine tools; having complete chapters on the lathe, the circular saw and the band saw. It gives many examples of pattern work, each one fully illustrated and explained with much detail. These examples, in their great variety, offer much that will be found of interest to all pattern makers, and especially to the younger ones, who are seeking information on the more advanced branches of their trade. Containing nearly 350 pages and 170 illustrations. Second edition, revised and enlarged. **Price, $2.50**

PERFUMERY

Henley's Twentieth Century Book of Receipts, Formulas and Processes.

Edited by G. D. HISCOX. The most valuable techno-chemical receipt book published. Contains over 10,000 practical receipts, many of which will prove of special value to the perfumer. **Price, $4.00**

Perfumes and Cosmetics, Their Preparation and Manufacture.

By G. W. ASKINSON, Perfumer. A comprehensive treatise, in which there has been nothing omitted that could be of value to the perfumer or manufacturer of toilet preparations. Complete directions for making handkerchief perfumes, smelling-salts, sachets, fumigating pastiles; preparations for the care of the skin, the mouth, the hair, cosmetics, hair dyes and other toilet articles are given, also a detailed description of aromatic substances; their nature, tests of purity, and wholesale manufacture, including a chapter on synthetic products, with formulas for their use. A book of general, as well as professional interest, meeting the wants not only of the druggist and perfume manufacturer, but also of the general public. Fourth edition much enlarged and brought up-to-date. Nearly 400 pages, illustrated. **$5.00**

PLUMBING

Standard Practical Plumbing.

By R. M. STARBUCK. This is a complete treatise and covers the subject of modern plumbing in all its branches. It treats exhaustively on the skilled work of the plumber and the theory underlying plumbing devices and operations, and commends itself at once to anyone working in any branch of the plumbing trade. A large amount of space is devoted to a very complete and practical treatment of the subjects of hot water supply, circulation and range boiler work. Another valuable feature is the special chapter on drawing for plumbers. The illustrations, of which there are three hundred and forty-seven, one hundred being full-page plates, were drawn expressly for this book and show the most modern and best American practice in plumbing construction. 6½ x 9¼. Cloth, 406 pages, 347 illustrations. **Price, $3.50**

Mechanical Drawing for Plumbers.

By R. M. STARBUCK. A concise, comprehensive and practical treatise on the subject of mechanical drawing in its various modern applications to the work of all who are in any way connected with the plumbing trade. Nothing will so help the plumber in estimating and in explaining work to customers and workmen as a knowledge of drawing, and to the workman it is of inestimable value if he is to rise above his position to positions of greater responsibility. 150 illustrations. **Price, $2.00**

Modern Plumbing Illustrated.

By R. M. Starbuck. The author of this book, Mr. R. M. Starbuck, is one of the leading authorities on plumbing in the United States. The book represents the highest standard of plumbing work. A very comprehensive work, illustrating and describing the drainage and ventilation of dwellings, apartments and public buildings. The very latest and most approved methods in all branches of sanitary installation are given. The standard book for master plumbers, architects, builders, plumbing inspectors, boards of health, boards of plumbing examiners and for the property owner, as well as the workman and apprentice. It contains fifty-seven entirely new and large full pages of illustrations with descriptive text, all of which have been made specially for this work. These plates show all kinds of modern plumbing work. Each plate is accompanied by several pages of text, giving notes and practical suggestions, sizes of pipe, proper measurements for setting up work, etc. Suggestions on estimating plumbing construction are also included. 407 octavo pages, fully illustrated by 57 full-page engravings. **Price, $5.00**

RECIPE BOOK

Henley's Twentieth Century Book of Recipes, Formulas and Processes.

Edited by Gardner D. Hiscox. The most valuable techno-chemical formulæ book published, including over 10,000 selected scientific, chemical, technological and practical recipes and processes. This book of 800 pages is the most complete book of recipes ever published, giving thousands of recipes for the manufacture of valuable articles for everyday use. Hints, helps, practical ideas and secret processes are revealed within its pages. It covers every branch of the useful arts and tells thousands of ways of making money and is just the book everyone should have at his command. The pages are filled with matters of intense interest and immeasurable practical value to the photographer, the perfumer, the painter, the manufacturer of glues, pastes, cements and mucilages, the physician, the druggist, the electrician, the dentist, the engineer, the foundryman, the machinist, the potter, the tanner, the confectioner, the chiropodist, the manufacturer of chemical novelties and toilet preparations, the dyer, the electroplater, the enameler, the engraver, the provisioner, the glass worker, the goldbeater, the watchmaker and jeweler, the ink manufacturer, the optician, the farmer, the dairyman, the paper maker, the metal worker, the soap maker, the veterinary surgeon, and the technologist in general. A book to which you may turn with confidence that you will find what you are looking for. A mine of information up-to-date in every respect. Contains an immense number of formulas that every one ought to have that are not found in any other work. 1920 edition. Cloth binding. **Price, $4.00**

RUBBER

Henley's Twentieth Century Book of Receipts, Formulas and Processes.

Edited by GARDNER D. HISCOX. Contains upward of 10,000 practical receipts, including among them formulas on artificial rubber. **Price, $4.00**

Rubber Hand Stamps and the Manipulation of India Rubber.

By T. O'CONOR SLOANE. This book gives full details of all points, treating in a concise and simple manner the elements of nearly everything it is necessary to understand for a commencement in any branch of the India rubber manufacture. The making of all kinds of rubber hand stamps, small articles of India rubber, U. S. Government composition, dating hand stamps, the manipulation of sheet rubber, toy balloons, India rubber solutions, cements, blackings, renovating varnish, and treatment for India rubber shoes, etc.; the hektograph stamp inks, and miscellaneous notes, with a short account of the discovery, collection and manufacture of India rubber are set forth in a manner designed to be readily understood, the explanation being plain and simple. Third edition. 175 pages, illustrated. **Price, $1.25**

SAWS

Saw Filing and Management of Saws.

By ROBERT GRIMSHAW. A practical hand book on filing, gumming, swaging, hammering and the brazing of band saws, the speed, work, and power to run circular saws, etc. A handy book for those who have charge of saws, or for those mechanics who do their own filing, as it deals with the proper shape and pitches of saw teeth of all kinds and gives many useful hints and rules for gumming, setting, and filing, and is a practical aid to those who use saws for any purpose. Third edition, revised and enlarged. Illustrated. **Price, $1.50**

SCREW CUTTING

Threads and Thread Cutting.

By COLVIN and STABLE. This clears up many of the mysteries of thread cutting, such as double and triple threads, internal threads, catching threads, use of hobs, etc. Contains a lot of useful hints and several tables. Third edition. **35 cents**

STEAM ENGINEERING

Horse-power Chart.

Shows the horse-power of any stationary engine without calculation. No matter what the cylinder diameter or stroke; the steam pressure or cut-off; the revolutions, or whether condensing or non-condensing, it's all there. Easy to use, accurate, and saves time and calculations. Especially useful to engineers and designers. **Price, 50 cents**

Steam Engine Troubles.

By H. Hamkens. It is safe to say that no book has ever been published which gives the practical engineer such valuable and comprehensive information on steam engine design and troubles. There are descriptions of cylinders, valves, pistons, frames, pillow blocks and other bearings, connecting rods, wristplates, dashpots, reachrods, valve gears, governors, piping, throttle and emergency valves, safety stops, flywheels, oilers, etc. If there is any trouble with these parts, the book gives you the reasons and tells how to remedy them. 350 pages, 276 illustrations. **Price, $2.50**

American Stationary Engineering.

By W. E. Crane. A new book by a well-known author. Begins at the boiler room and takes in the whole power plant. Contains the result of years of practical experience in all sorts of engine rooms and gives exact information that cannot be found elsewhere. It's plain enough for practical men and yet of value to those high in the profession. Has a complete examination for a license. Third edition revised and enlarged. 345 pages. 131 illustrations. Cloth. **Price, $2.50**

Steam Engine Catechism.

By Robert Grimshaw. This volume of 413 pages is not only a catechism on the question and answer principle, but it contains formulas and worked-out answers for all the steam problems that appertain to the operation and management of the steam engine. Sixteenth edition. **Price, $2.00**

Boiler Room Chart.

By Geo. L. Fowler. A chart—size 14 x 28 inches—showing in isometric perspective the mechanism belonging in a modern boiler room. The various parts are shown broken or removed, so that the internal construction is fully illustrated. Each part is given a reference number, and these, with the corresponding name, are given in a glossary printed at the sides. **Price, 25 cents**

Engine Runner's Catechism.

By Robert Grimshaw. Tells how to erect, adjust and run the principal steam engines in use in the United States. The work is of a handy size for the pocket. To young engineers this catechism will be of great value, especially to those who may be preparing to go forward to be examined for certificates of competency; and to engineers generally it will be of no little service, as they will find in this volume more really practical and useful information than is to be found anywhere else within a like compass. 387 pages. Seventh edition. **Price, $2.00**

Modern Steam Engineering in Theory and Practice.

By Gardner D. Hiscox. This is a complete and practical work issued for stationary engineers and firemen dealing with the care and management of boilers, engines, pumps, superheated steam, refrigerating machinery, dynamos, motors, elevators, air compressors, and all other branches with which the modern engineer must be familiar. Nearly 200 questions with their answers on steam and electrical engineering, likely to be asked by the examining board, are included. Third edition. 487 pages, 405 engravings. Cloth. **Price, $3.50**

Steam Engineer's Arithmetic.

By Colvin-Cheney. A practical pocket book for the steam engineer. Shows how to work the problems of the engine room and shows "why." Tells how to figure horse-power of engines and boilers; area of boilers; has tables of areas and circumferences; steam tables; has a dictionary of engineering terms. Puts you onto all of the little kinks in figuring whatever there is to figure around a power plant. Tells you about the heat unit; absolute zero; adiabatic expansion; duty of engines; factor of safety; and 1,001 other things; and everything is plain and simple—not the hardest way to figure, but the easiest. Second edition. **Price, 75 cents**

STEAM HEATING and VENTILATING

Practical Steam, Hot-Water Heating and Ventilation.

By A. G. King. This book has been prepared for the use of all engaged in the business of steam, hot-water heating and ventilation. Tells how to get heating contracts, how to install heating and ventilating apparatus, the best business methods to be used, with "Tricks of the Trade" for shop use. Rules and data for estimating radiation and cost and such tables and information as make it an indispensable work for everyone interested in steam, hot-water heating and ventilation. It describes all the principal systems of steam, hot-water, vacuum, vapor and vacuum-vapor heating, together with the new accelerated systems of hot-water circulation, including chapters on up-to-date methods of ventilation and the fan or blower system of heating and ventilation. Second edition. 367 pages, 300 detailed engravings. Cloth. **$3.50**

500 Plain Answers to Direct Questions on Steam, Hot-Water, Vapor and Vacuum Heating Practice.

By Alfred G. King. This work, just off the press, is arranged in question and answer form; it is intended as a guide and text-book for the younger inexperienced fitter and as a reference book for all fitters. All long and tedious discussions and descriptions formerly considered so important have been eliminated, and the theory and laws of heat and the various old and modern methods and appliances used for heating and ventilating are treated in a concise manner. This is the standard Question and Answer examination book on Steam and Hot Water Heating, etc. 200 pages, 127 illustrations. Octavo. Cloth. **Price, $2.00**

STEEL

Hardening, Tempering, Annealing and Forging of Steel.

By J. V. Woodworth. A book containing special directions for the successful hardening and tempering of all steel tools. Milling cutters, taps, thread dies, reamers, both solid and shell, hollow mills, punches and dies, and all kinds of sheet-metal working tools, shear blades, saws, fine cutlery and metal-cutting tools of all descriptions, as well as for all implements of steel, both large and small, the simplest, and most satisfactory hardening and tempering processes are presented. 320 pages, 250 illustrations. Fourth edition. Cloth. **Price, $3.00**

Steel: Its Selection, Annealing, Hardening and Tempering.

By E. R. Markham. This work was formerly known as "The American Steel Worker," but on the publication of the new, revised edition, the publishers deemed it advisable to change its title to a more suitable one. This is the standard work on hardening, tempering, and annealing steel of all kinds. This book tells how to select, and how to work, temper, harden, and anneal steel for everything on earth. It is the standard book on selecting, hardening, and tempering all grades of steel. 400 pages. Very fully illustrated. Fourth edition. **Price, $3.00**

TRACTORS

The Modern Gas Tractor.

By Victor W. Pagé. A complete treatise describing all types and sizes of gasoline, kerosene, and oil tractors. Considers design and construction exhaustively, gives complete instruction for care, operation and repair, outlines all practical applications on the road and in the field. The best and latest work on farm tractors and tractor power plants. A work needed by farmers, students, blacksmiths, mechanics, salesmen, implement dealers, designers, and engineers. Second edition revised and much enlarged. 504 pages. Nearly 300 illustrations and folding plates. **Price, $3.00**

TURBINES

Marine Steam Turbines.

By Dr. G. Bauer and O. Lasche. Assisted by E. Ludwig and H. Vogel. Translated from the German and edited by M. G. S. Swallow. The book is essentially practical and discusses turbines in which the full expansion of steam passes through a number of separate turbines arranged for driving two or more shafts, as in the Parsons system, and turbines in which the complete expansion of steam from inlet to exhaust pressure occurs in a turbine on one shaft, as in the case of the Curtis machines. It will enable a designer to carry out all the ordinary calculation necessary for the construction of steam turbines, hence it fills a want which is hardly met by larger and more theoretical works. Numerous tables, curves and diagrams will be found, which explain with remarkable lucidity the reason why turbine blades are designed as they are, the course which steam takes through turbines of various types, the thermodynamics of steam turbine calculation, the influence of vacuum on steam consumption of steam turbines, etc. In a word, the very information which a designer and builder of steam turbines most requires. Large octavo, 214 pages. Fully illustrated and containing 18 tables, including an entropy chart. **Price, $4.00 net**

(See page 35 for further description of the book.)